Transition Engineering

Building a Sustainable Future

Susan Krumdieck

University of Canterbury, New Zealand

CRC Press
Taylor & Francis Group
Boca Raton London New York

CRC Press is an imprint of the
Taylor & Francis Group, an **informa** business

CRC Press
Taylor & Francis Group
6000 Broken Sound Parkway NW, Suite 300
Boca Raton, FL 33487-2742

© 2020 by Taylor & Francis Group, LLC
CRC Press is an imprint of Taylor & Francis Group, an Informa business

No claim to original U.S. Government works

Printed on acid-free paper

International Standard Book Number-13: 978-0-367-34126-8 (Paperback)
978-0-367-36243-0 (Hardback)

Visit the Taylor & Francis Web site at
http://www.taylorandfrancis.com

and the CRC Press Web site at
http://www.crcpress.com

Dedication

This book is dedicated to the memory and life's work of Professor Frank Kreith, who was a child-survivor of the Nazi holocaust. Professor Kreith immigrated to the United States and became a mechanical engineering professor. He made immense contributions to the fields of heat transfer and solar energy. Professor Kreith was active in the sustainable energy area, and he was passionate about the prospects for the energy transition. I met Professor Kreith in 2011 at a dinner party in Boulder, Colorado. We started a discussion about sustainable energy that lasted for 8 years. He felt that politics and economics had somehow diverted the workforce of change and that the promise of sustainable energy was being lost. He understood the premise that the change of perspective in the engineering professions described in Transition Engineering: Building a Sustainable Future *could be the key, the missing component of the energy transition. This energized me to put these ideas together in a way that could be quickly learned and put into action. I started exploring this Transition Engineering perspective about 30 years ago. My students and colleagues have demonstrated how this approach works. But it wasn't until Professor Kreith said, 'I think you have found the key!' that I started to think of Transition Engineering as an inevitable evolution of our professional engineering capabilities and started teaching about it.*

Contents

Foreword

Transition Engineering is emerging in response to the mega-issues of global climate change, the inevitable and necessary decline in the world's supply of oil, scarcity of key industrial minerals and local environmental constraints. These sustainability issues underpin today's so-called wicked problems and pose strategy challenges for organizations, businesses and communities. The Transition Engineering framework presented in *Transition Engineering: Building a Sustainable Future* is a methodology for working on wicked problems. It is also a methodology for communities and organizations to explore big shifts in development away from unsustainable activities.

World consumption and population growth have been accompanied by growth in total energy supply and by improvements in productivity and efficiency. More than 90% of the current energy supply is from fossil hydrocarbons, finite resources that produce a potent greenhouse gas, CO_2. Resource, environmental and social limits to growth have been studied since the 1970s, and a wide range of indicators show that the factors that drove growth in the previous century are slowing (Stern 2006). The energy transition is the only realistic approach to mitigating the most destructive climate impacts of increased greenhouse gas concentrations (Lord 1955). Transition Engineering projects are about changing existing complex systems to radically lower energy and material use while preserving essential functions. It is challenging work. In fact, it is so hard that most of the problems we will examine are referred to as wicked problems. A wicked problem has no solution, no answer and no possibility of successful outcomes if current conditions continue. Wicked problems are usually seen as complex social and economic problems, but there are usually engineered systems at the heart of the issues.

We cannot predict the future. However, we know that development in this century will be different from the last, and the fundamental problem will be energy. Engineers in all disciplines will be responsible for accomplishing the energy transition, which most analysts agree cannot happen without substantial demand reduction, conservation and curtailment (Meadows 2004). Demand reduction means that consumers choose to use less energy and fewer energy-intensive services or products because of price. A carbon tax, emission trading or subsidies of renewables are strategies for effecting reduction of demand for fossil fuels. However, the lack of market availability of lower-cost alternatives to fossil fuels has meant that demand reduction has not reduced carbon emissions to date. Conservation means that the service is provided using a smaller amount of energy. Energy efficiency improvements for vehicles, power generation and manufacturing are the main contributors. Curtailment means that a service or product is no longer used.

Transition Engineering involves the work of analyzing energy development history, using science-based scenarios; creative brainstorming; idea generation and innovation for designing, planning and implementing major change projects that eliminate fossil fuel use in a particular existing system in a specific location. Transition Engineering projects involve companies, organizations and communities.

Each new project aims to realize opportunities in change. Transition Engineers use strategic problem solving to carry out projects that develop resilience and adaptive capacity. Transition Engineering projects must have technical and economic viability in the near term. These projects aim to produce multiple benefits in the medium term and to achieve long-term environmental, social, economic and cultural wellbeing. All successful Transition Engineering projects will improve the competitive position of the organization now, each year through the payback period, and beyond the project lifetime.

ENGINEERING APPROACH

Engineering is the art of applying knowledge and science to solve problems and make things work. All engineering fields agree on the following basic rules of engineering:

1. Know what problem you are working on.
2. Define the system and relationships to other systems.
3. Identify what is essential and what is optional.
4. Draw a picture.
5. Keep it simple.
6. Borrow from the best, learn from the rest.
7. Is it worth it?
8. Learn from mistakes.

Before the turn of the twentieth century, all growth in energy use was from coal. After the Great Depression and World War II, the idea that growth was necessary for increased wealth of a nation took hold. For more than two generations, the culture of growth has prevailed and formed the basis for decision-making and policy. For more than 50 years, the issue of carbon dioxide accumulation in the atmosphere and a whole host of pollution and depletion issues have been well known.

In Chapter 1, we will discuss the mega-issues of global warming and peak oil. Chapter 2 is a survey of the wide-ranging problems of unsustainable energy. Chapter 3 looks at how we communicate about how energy systems work, what we understand about the past and how we look at the future. The interdisciplinary transition innovation, management and engineering (InTIME) approach set out in Chapter 4 is a new approach that refocuses our perception directly onto the main problem and helps us generate new ideas. Every new idea requires investment, but energy transition investment will require new perspectives, and some of the fundamental concepts and analysis tools used in the InTIME approach are explained in Chapter 5. Chapter 6 provides instruction in conventional economic analysis used in engineering projects. Chapter 7 explores several new ways to communicate the opportunities and competitive advantages of investing in the energy transition and especially the benefits of being innovators and early adopters. The decline in the world's supply of oil and gas is essentially inevitable because of the finite nature of the resource and the extraction processes. In the concluding chapter, Chapter 8, the ultimate shift project is presented for the ultimate wicked problem – oil.

Transition Engineering: Building a Sustainable Future should be of use for engineers from all disciplines. The Transition Engineering methods and analyses are covered in depth, and the system-level aspects of the different disciplines are treated in a way that is accessible to engineers in different fields.

The one thing that readers should notice about *Transition Engineering: Building a Sustainable Future* that is different from other sustainable energy texts is that I will not use the phrase 'we must' and I will not call on any politicians, economists, companies or consumers to act. This book is about the engineering work of the energy transition. It is about what engineers will be able to do and how we will do it.

Introduction

Can you change the future?

Let's say that I have a time machine and I just sent you back in time to the morning of 14 August 1912. It is dawn, and you are on the deck of a huge ship, in fact, the largest and most luxurious passenger ship ever built, and it is a bit cold in the North Atlantic, even though the sea is like glass. I will bring you back on 17 August, if you make it to New York. If you don't make it to New York, you will be one of the more than 1,500 casualties of the sinking of the *Titanic*. All you have to do is solve the problem of how to change the future and save the *Titanic*.

You have perfect knowledge of the future (Lord 1955). This type of certainty about future scenarios is the thing that problem solvers and sustainability advocates wish they had so that they could convince politicians, economists and consumers to take action on climate change. You know that in about 16 hours the lookout will raise the alarm about an iceberg right ahead, but that warning will be entirely too late as the *Titanic* strikes the massive iceberg at full speed, less than two minutes after it is sighted. You know that the captain, crew and everyone on board believe they are on an 'unsinkable ship' with advanced design of six watertight bulkhead compartments. You know there will be catastrophic failure of those compartments, and you are the one person on that ship who is certain that hitting the iceberg at full speed will sink the ship – and it will sink spectacularly fast.

There is a 30% chance you will make it to New York if you manage to get into one of the lifeboats. You know that the lifeboat capacity is enough for only about half of the people on board but that around 500 places will remain unfilled. This is largely because the crew have never done evacuation drills, and the captain will order that women and children go first, which causes chaos as families resist separation and women refuse to go. Thus, your first thought of how to use your perfect foresight is to make sure that you are in the right place and to dress like a woman travelling first-class. You could position yourself for survival by spending the remaining hours finding a disguise and locating lifeboat number 1, which will be lowered with only a few people in it. But you still have some time, and the real problem is that the ship will hit an iceberg at full speed. Thus, the real solution is *not* to hit the iceberg. What else could you do? You now have 12 hours until 11:40 p.m.

You know that Captain Smith and the company director, Mr. Ismay, will receive the information that there are large icebergs and ice floes directly in the path of the ship. The two wireless operators on the *Titanic*, Harold and Jack, will receive these messages from at least six other ships in the area throughout the day. The *Titanic* has the most powerful wireless ever made, and it has been widely advertised that messages can be sent to New York during the whole trip. Harold and Jack are swamped with telegraph messages from first-class passengers to their friends reporting the great time they are having, the opulence of the ballroom, and who they met. At some point later in the evening, you know Captain Smith will instruct a slight course change to the south, believing that this is a sufficient response. He will give the orders to Second Officer Lightoller to maintain full speed unless fog arises, as he goes off-duty to have dinner with first-class passengers. At 10:40 p.m., Jack will record an urgent message from another liner giving the location of an unusually large ice floe with coordinates directly in the path of the *Titanic*. Jack would not have known that the coordinates matched the navigation route. It has often been speculated that this last and most relevant message, which could have saved the ship, was not relayed to the bridge. Of course, from the historical perspective, the *first* warning could have saved the ship if Captain Smith had ordered reduced speed.

Just like the thousands of climate scientists in our present time, the telegraph operators on the *Titanic* actually have the information that should instigate the action to avert disaster. You decide to go tell them that they need to find better ways to communicate, and even if they are very busy with all their other work, they must get the messages to the bridge. It takes you a while to find where the telegraph operators actually are on the massive ship. You explain how important the messages are about the icebergs. They tell you that they have already passed on the messages and done their jobs. They tell you they can't do anything about the way the ship is operated, and don't know anything about coordinates, so go talk to the navigator!

Just like the economists of our day who chart the course of the economy, the navigator's job is to chart the course that keeps the *Titanic* going at full speed. The maiden voyage of the *Titanic* was advertised to be the most luxurious and fastest crossing of the Atlantic. The navigator is confident that he knows the way. The route for this fastest trip is already decided. "Look at the chart here – see the curve?" The navigator says that it's his job to plot the course for maximum speed, and he wouldn't even know how to slow down the ship at any rate. Go talk to the captain!

Just like the politicians of our day who keep the economy going according to plan, the captain's job is to make the key decisions that keep the *Titanic* sailing according to plan. He is also part of a profit-generating business and must work to meet the objectives of the White Star Line. You have a very difficult time getting near the bridge, so you see Captain Smith down the hall as he returns from dinner. You yell out to him that the ship is at risk and ask him to order a reduction in speed. He doesn't seem receptive to your warning and assures you that he has hit icebergs before so not to worry. The *Titanic* has the latest technology and is too big to sink. Besides, he can't slow down because the company would lose money. Go talk to the company director!

It is now 9:30 p.m. and it is getting late. You are running out of time. You try to find a way to talk to the director, Mr. Ismay, but you are not allowed into the first-class area. While you are hanging around the stateroom where the orchestra is playing, a first-class passenger comes down the hall. You try to talk to him about the risk facing the ship and urge him to ask the captain to slow down the ship and warn the director that his company could lose its investment. He seems like a nice guy, but he assures you that, having paid a premium for passage, he is confident that all will be well.

You decide to go below deck to talk to the lower-class passengers. You explain the facts: that the wireless operators have information about dangerous icebergs and you urge them to demand that the captain reduce the speed. A few of the young people agree to go with you to occupy the upper deck and protest. But as you try to return to the promenade, you find that it is now 10:30 p.m. and the gates separating the decks have been locked. Now you are really stuck because you can't go back up to get to the lifeboats. Your only chance is to find some way to save the *Titanic*!

Lucky for you, you studied thermodynamics at university. You recall that the power developed by a steam cycle is directly related to the rate of heat input. It is the energy system that is driving the ship to disaster. You know how to slow down the ship!

You rush to the engine room in the bowels of the ship to find the 25 members of the engineering staff, all of whom you know will perish in the sinking. You find the firemen shovelling fuel into the boiler as fast as possible. You explain about the information from other ships in the area warning of huge icebergs directly in the ship's path. The engineers say that they have not received any order to reduce speed, and they would get in trouble if they slowed down the consumption rate of coal. Looking at your watch, you can assure them that it will be such a close call that everyone will be able to see the danger and will be glad they had the safety margin to steer around the deadly obstacle. But the engineers are reluctant to take action on their own, and they remind you that the new bulkhead design means that the ship is unsinkable. They work for the company and must just do their jobs. You ask that if the company were really responsible, would it have allowed the *Titanic* to sail with lifeboats for only half the passengers and crew? And you remind them that the company has been losing the competition for first-class passengers and that the directors had decided to send the ship out into the Atlantic with a smouldering fire in a coalbunker, the heat of which has weakened the bulkhead. The great ship is now susceptible and the myth of being unsinkable is busted. You point out that they, as engineers, have the ultimate responsibility for all of the passengers' safety and that, as engineers, they should know that it is folly to believe in infallibility. They are working men, not ship owners or captains. If you could convince them to stick together, take responsibility for the lives of all the other people on board, and go against the wishes of the owners and the orders from the bridge, then you can save the *Titanic*.

Your life and the future of the ship could have been secured by the scientific information being communicated, the company being responsible, the leadership being precautionary and/or the passengers taking action. But now the future depends on the 1% of the people on board who are actually making the ship go full steam ahead – directly into catastrophe.

Can you convince the engineers to slow down the consumption rate of coal, power down the steam engines, and save the *Titanic*?

Acknowledgements

Transition Engineering is an emerging field. It is emerging out of the disciplines of sustainability and renewable energy. Thus, I wish to acknowledge the shoulders of the giants we stand on, particularly Professor Frank Kreith. I sincerely thank my colleagues who jumped off the deep end with me to boldly explore the post oil transition: Andre Dantas, Elijah van Houten, Femke Reitsma, Rua Murray, Raphael Boichot, and Anne-Catherine Favre-Pugin. I wish to acknowledge the contributions of my PhD, graduate and undergraduate students, who have been part of the Advanced Energy and Material Systems Lab (AEMSLab) and my Energy Transition Engineering course at Canterbury University in New Zealand: Shannon Page, Hadley Cave, Michael Saunders, Stacy Rendall, Michael Southon, Leighton Taylor, Jake Frye, Andy Hamm, Montira Watcharasukarn, Sam Gyamfi, Mik Dale, Janice Asuncion, Miraz Fulhu, Muavi Mohammed, Ming Bai, Neibert Blair, Isabel Gallardo, Patricio Gallardo, Jeremy Pascal and Johann Land. The foundation of the Global Association for Transition Engineering (GATE) has been a huge milestone and was the vision of Daniel Kenning, whose professional engineering approach has contributed greatly to this text.

In particular I wish to acknowledge the trigger for this whole field, my son, Kyan Krum dieek. More than a decade ago, after watching "An Inconvenient Truth" he asked me if climate change was really as bad as it seemed. Of course, I answered truthfully, that it is in fact worse than portrayed in the movie. He then asked if the sustainable energy technologies that I taught in my class and focused my research on would fix global warming. It was devastating, but I had to answer factually, that no; even if all the sustainable technologies were developed as fast as possible, it would not be enough. My son then said, "OK, Mom, you have to do something!".

About the Author

Susan Krumdieck is a professor in the Department of Mechanical Engineering at the University of Canterbury in New Zealand, where she has taught energy Transition Engineering for 17 years. Her research focuses on developing the engineering methods and innovative technologies for adaptation to down-shift fossil fuel production and consumption. She is an expert in developing new ideas for achieving decarbonization in transportation systems and urban regeneration. Her visionary contributions involve reimagining the urban built environment and transportation activity systems by developing new Geographical Information System (GIS) and modelling methods that combine diverse aspects of urban systems. Her research group has produced groundbreaking ideas about engineering development policy and community vision by communicating in new ways about history and the long-term future. Their research asks, 'Given that 80% reduction in greenhouse gas is required, the context of complex problems of unsustainability can be defined, and the value of transition to something better is understood, what is the next practical and transformative project for a given city or company?'

Professor Krumdieck is the leader of a group of engineering professionals and academics establishing the field of Transition Engineering. She is the co-founder and a trustee of the Global Association for Transition Engineering (GATE). Transition Engineering is the work of designing and carrying out change projects for industry and the public sector that meet COP21 targets while providing multiple short- and long-term benefits. She was appointed to the Royal Society of New Zealand (RSNZ) Energy Panel in 2005 and contributed to long-term energy transition analysis and the role of demand-side innovations. She was selected as the Institute of Engineering and Technology (IET) prestige lecturer in 2010 and presented the lecture on Transition Engineering around New Zealand and in London and Brussels. She won the CU Gold Sustainability Award in 2011 for organizing *Signs of Change*, the first national no-travel conference, and the Silver Sustainability award in 2013 for her work on From the Ground Up, an urban redevelopment approach. She has worked with Professor Frank Kreith on the first energy engineering text that included the topic of Transition Engineering.

Professor Krumdieck was the recipient of a fellowship from the Scientific Council of Grenoble Institut Polytechnique (INP) for research and teaching, and she delivered a well-received course on Transition Engineering in 2015. She also received a fellowship to teach a Masters course on Transition Engineering at the University of Duisburg-Essen in 2015. In 2016 and 2017, Susan was a guest of Bristol University, where she presented workshops on Transition Engineering and consulted with an academic working group to design a curriculum for a Transition Engineering masters degree program. Susan was an Honorary Fellow of Munich University of Applied Sciences in 2018 to engage in research collaboration and to teach a semester course on Transition Engineering.

Professor Krumdieck serves on the editorial board for six journals, including *Energies, Energy Conservation & Management* and *Biophysical Economics*, and she has edited special issues of *Energy Policy, Energies*, and *Sustainability*. She has been an invited participant in the New Zealand Climate Forum, Ministry roundtables and workshops on future transport and energy strategies, and she currently serves on the University of Canterbury (UC) Sustainability Panel, and the Ministry of Transport Upper North Island Freight Supply Chain Strategy Working Group. She serves on the Scientific Advisory Board, Regional Center for Energy and Environment Sustainability (RCEES) Africa Centers of Excellence.

Robert Stannard is an animator, illustrator and sometimes actor. He loves bringing stories to life whether that be through film or still images and is working in Auckland, New Zealand, as a 3D animator for TV and Film.

1 The Mega-Problems of Unsustainability

Know what problem you are working on.

1.1 INTRODUCTION: THE MEGA-PROBLEMS

Engineers learn the methodology for solving problems during their studies at university. First you define the system and then state what must be found, and above all – know what problem you are trying to solve. Time is wasted, and creativity stifled, if you try to jump to a solution. But it is so tempting to over-simplify the problem and jump to a solution. Simple, sustainable energy solutions are easily found in a search on the internet. Is the problem that people aren't using enough sustainable energy, or is the problem that they are using too much unsustainable energy?

There are no simple solutions to the complex problems of unsustainable energy. Energy transition is the down-shift of fossil fuel use to 80% below 2009 levels (WFEO 2015). Transition Engineering is the emerging field that achieves the thousands of down-shift projects that change existing systems. The main approach and methods for Transition Engineering are presented in Chapter 3. In this chapter, we learn about the problems of unsustainability and how to deal with our desire for simple solutions. We will consider all types of activity systems that consume fossil fuel, as well as the energy supply systems. The energy transition perspective includes technologies, but you will also learn how to understand the social, economic and environmental contexts around energy supply and end use. Probably the most challenging part of the Transition Engineering approach is the ability to think about current development options, while also engineering for the long-term future.

Sustainability has been a stated goal of most organizations and businesses for more than 30 years. The Internet is full of green technology stories attracting thousands of Facebook shares and likes. It would be irrational for a politician in any party to say that he or she is fighting against sustainability. Business leaders would not call for the elimination of sustainability. We all agree we want sustainability, yet virtually all measures of resource depletion and environmental quality around the world have been getting worse (Worldwatch Institute 1984–2016).

1.1.1 THE MEGA-PROBLEMS: GREENHOUSE GAS ACCUMULATION AND OIL SUPPLY

There are two mega-challenges for energy systems: the global warming caused by combustion of fossil carbon, and the social and economic issues of oil supply. The term *mega-challenges* (Winston 2014) is used because all sectors of the economies and societies of all nations are affected on a large scale for an undeterminable time into the future, and the effects are not positive. Energy transition will be the change in energy supply, economics and end uses achieving deep de-carbonization.

Benefits from change are more likely to be realized if the challenges are understood and changes planned before a crisis occurs (Brown 2010). Unplanned changes usually have a high economic and social cost. Many people find these mega-challenges difficult to deal with psychologically. But there is an upside: innovation can occur only when we are challenged by the lack of known solutions. The risks of climate change and energy security are serious, but don't get stuck. Use this chapter to gain a good understanding and come to grips with the problems so we can move on to what we are going to do about them. These mega-challenges are huge complex subjects. You are encouraged to learn more about them through your own study. The treatment here will focus on getting to the core engineering requirements for Transition Engineering projects.

1.2 THE PROBLEM WITH SUSTAINABLE DEVELOPMENT: IT ISN'T WORKING

I entered engineering at university in 1981. I was passionate about sustainability and especially about sustainable energy. The Four Corners Power Plant had been built in the previous decade and had caused serious pollution and acidification of mountain lakes and streams in the remote and unpopulated wilderness of its surrounding region. I wanted to figure out how to fix polluting cars and power plants and how to change to renewable energy. Hydroelectricity was already a main source of power at that time, but the cost in loss of beautiful landscapes had been high. Wind turbines, solar hot water and Photovoltaic (PV) panels, and geothermal power plants are much more reliable, larger scale and lower cost than they were in 1981. But I would like young readers to know that 30 years ago we were told that if we worked on these alternatives, including electric cars, fuel cells and biofuels, to make them competitive with fossil fuels, then we would see the transition to sustainable energy. There is something fundamentally wrong with that story. The reason that the field of Transition Engineering now exists is because reducing fossil fuel production and use is a major engineering challenge; it will not be the natural result of achieving viability of renewable alternatives. It is fine and good that renewable energy technologies and energy efficiency have progressed substantially since the 1980s, but it is not sufficient.

1.2.1 A SHORT WORLD HISTORY OF DEVELOPMENT

Unsustainability is the fundamental problem. Civilizations are either sustainable or they are not, and their worldview plays a part in whether they change unsustainable activities (Diamond 2005). In the ancient world, the worldview reflected the cyclical pattern of life. Starting in the classical Greco-Roman period, a more linear view of time emerged. By the Middle Ages, it was normal to think that society would cumulatively progress from lesser or more primitive forms to more developed and moral forms, with the ultimate spiritual perfection being found in the next world.

Through the eighteenth and nineteenth centuries, science became inextricably linked with the concept of an overall plan for the march of humankind that involved mastery over nature and developing the wilderness for the betterment of civilization. The industrial revolution provided the proof of the doctrine of progress and the rightness of humanity's domination of nature. By the beginning of the current century, Americans and Europeans had a widespread belief in progress and that things only have value if they are produced by industry and given license by the magic of the market (Du Pisani 2006).

It is important for you to recognize your belief in the unstoppable progress of humankind through the optimism of science and technology, and the virtue of the freedom of the individual to capitalize and profit from this march of inevitable development. Look at a graph of any indicator of industrial progress, such as the production of oil or iron ore, kilometres travelled or cars manufactured, and you will see a sharp exponential trend upwards starting after World War II. The population was growing fast. The growth in pollution and environmental impacts from resource extraction and manufacturing were so rapid that the degradation of land, air and water caused widespread alarm. This was not the first time that human activity had caused environmental degradation that threatened prosperity. In the fifth century BC, the philosopher Plato warned about the loss of soil fertility and desertification of the forests due to excessive logging and farming.

The first use of the word *sustainability* was in 1713 in relation to the problem of deforestation in Germany. The term *nachhaltende Nutzung* referred to the sustainable use of the forests by balancing the harvest of old trees and regeneration of young trees. All across Europe the ancient forests were being rapidly depleted in the eighteenth century to build ships and growing cities and to fuel iron smelters and industry. The idea emerged in Germany to establish *ewiger Wald*, eternal forests that required aggressive afforestation and regeneration.

Great Britain found a different solution to the problem of forest depletion: the Crown's colonies in America, Australia, South America and New Zealand sustained the growth in shipbuilding and construction. After 1790 Great Britain's coal production increased rapidly, providing high-temperature heat energy to make steel and run steam engines. By the 1860s, the easy-to-access coal deposits were becoming depleted, and alarms were raised calling for more efficient and less wasteful use of coal (Du Pisani 2006). Luckily in this same timeframe, the science of thermodynamics and the practice of mechanical engineering were beginning to deliver significant improvements in efficiency. The progress of science and technology, along with the trend for globalization, sustained development after Europe's most accessible resources were depleted. The progress narrative of developed countries is steeped in a long history of new technologies and new resources being found to overcome problems.

The word *economy* meant management of household resources or increased value until the twentieth century, when it has become a thing: *the economy*. Politics focused on growing the economy. The global economic growth rate was maintained at an unprecedented 5.6% for more than 20 years, from 1948 to 1971. Economists have been aware of the problems of unsustainability. The theory is that as a resource becomes scarce, new technologies would be introduced to replace the scarce resource. Putting a price on environmental damage, called *externalities*, can stimulate new mitigation technologies, resulting in more economic growth. The current worldview holds that even as we face the social and environmental challenges of progress, this stimulates even more growth. This is one reason why the measure of economic growth, the gross domestic product (GDP), includes all spending whether it was on good things like milk, unhealthy things like soft drinks or remedies like new medical treatments for diabetes and tooth decay.

In recent decades, the increasing wealth gap between the developed and developing post-colonial countries gave rise to the idea that these countries must modernize. Modernization means opening their markets and providing incentives for international corporations to export their primary products and privatize their public services. By the 1980s, it was clear that the international wealth created by modernization was actually impoverishing the developing countries. The environmental movement in the 1970s saw the creation of non-governmental organizations such as Greenpeace and Friends of the Earth, whose activities were aimed at essentially stopping the most destructive development projects. After the energy and environmental crises of the 1960s and 1970s, a wave of books and films were released that extrapolated the growth rates in consumption and environmental damage and predicted collapse of population and industrial civilization (Meadows et al. 1972).

Since the 1970s, environmentalists and economists converged on the idea of sustainable development as the way to continue the project of civilization by steering technology into green products and green energy. Essentially, economists will not change their view, because it seems to be working fine, and environmentalists will not change their view, because the threats to the environment continue to get worse. Both economists and environmentalists also agree that wasteful consumer behaviour is one of the main problems, even though consumers do not have any control over the design of products, manufacturing processes, vehicles or buildings. If there is to be change, it appears it will be left to technologists to sort out what to change and how to accomplish it economically. History thus makes a good case for the emergence of Transition Engineering since society expects engineers to deliver the solutions to problems. The challenge in this century is that the project of progress involves dramatic reduction of energy and material consumption.

1.2.2 Sustainable Development

The most widely agreed upon concept of sustainable development is the *Brundtland Statement*. In 1983 the United Nations set up an independent body, the World Commission on Environment and Development, headed by Gro Harlem Brundtland, the former prime minister of Norway. The commission was asked to examine the critical environmental and developmental problems facing our planet and to formulate realistic proposals to solve them. The proposed solution was to make sure that human progress can be sustained, but without bankrupting the resources of future generations. The outcome of this study was an important book entitled *Our Common Future* (Brundtland 1987).

Brundtland Commission Definition of Sustainable Development (1987)
Sustainable development should meet the needs of the present without compromising the ability of future generations to meet their own needs.

Sustainable global development requires those who are more affluent to adopt lifestyles that can be accommodated by the planet's ecological means, particularly their use of energy. Further, rapidly growing populations increase the pressure on resources. Energy efficiency must be the primary vehicle of national energy strategies for sustainable development, but energy efficiency can only buy time for the world to develop energy paths based on renewable sources and this goal must be the foundation of any global energy strategy for the twenty-first century.

'o, the Brundtland Commission concluded that worldwide efforts to ...aintain human progress to meet human needs and to realize human ...ion are unsustainable, both in developed and developing nations, because they rely on ever-increasing use of already overdrawn environmental resources. Not included in the commission's report were the underpinning reasons for the problems in poor countries, the actions that should be taken, and who should take the actions.

Sustainable development is widely accepted as a motivational goal, but it has become clear that agreeing on a goal does not change what we are doing. In 2000, the United Nations General Assembly (UNGA) adopted the Millennium Declaration, which was signed by 189 countries, and set out eight millennium development goals (MDGs) with a 15-year agenda to end extreme poverty. The MDGs were aimed primarily at improving the human condition in developing countries, but none of the goals relate specifically to energy. In 2015 a new agenda on sustainable development was launched, which sets out 17 sustainable development goals (SDGs) and a list of 169 actions for individual citizens, companies and governments (UNGA 2015). Each goal has specific targets for the 15 years following 2015. The stated aim is to end poverty and hunger everywhere by 2030, and ensure human dignity and equality in peaceful, inclusive societies. The goals and actions are in reference to the principles of sustainable development: balancing the three dimensions of society, economy and environment.

The agenda does not specifically address energy supply in the aims, but there is a stated aim to 'protect the planet from degradation'. The aim is for 'sustainable consumption and production, sustainably managing natural resources, and taking urgent action on climate change'. The specific statements relating to energy are 'sustainable industrial development; universal access to affordable, reliable, sustainable and modern energy services; sustainable transport systems; and quality and resilient infrastructure', and 'reduction and recycling of waste and the more efficient use of water and energy'. The declaration also resolves to create conditions for sustainable economic growth and social and technical progress.

The key SDG for energy is Number 7 'Ensure access to affordable, reliable, sustainable and modern energy for all'. From an engineering perspective, the requirements for sustainable development are not easily quantifiable because of the words *accessible, affordable, reliable, sustainable* and *modern*. These are definitely attributes of the energy systems that most Europeans, Americans, Japanese and many Chinese enjoy. But these countries are using unsustainable energy systems. Here is a challenge: is the problem of sustainable development really about undeveloped countries or is it about transitioning developed countries?

United Nations Sustainable Development Goals (2015)
1. End poverty in all its forms everywhere.
2. End hunger, achieve food security and improved nutrition and promote sustainable agriculture.
3. Ensure healthy lives and promote well-being.
4. Ensure inclusive and equitable quality education and promote lifelong learning opportunities.

5. Achieve gender equality and empower women and girls.
6. Ensure availability and sustainable management of water and sanitation.
7. Ensure access to affordable, reliable, sustainable and modern energy for all.
8. Promote sustained, inclusive and sustainable economic growth and decent work.
9. Build resilient infrastructure, promote sustainable industrialization and foster innovation.
10. Reduce inequality within and among countries.
11. Make cities and human settlements inclusive, safe, resilient and sustainable.
12. Ensure sustainable consumption and production patterns.
13. Take urgent action to combat climate change and its impacts.
14. Conserve and sustainably use the oceans, seas and marine resources.
15. Protect, restore and promote sustainable use of terrestrial ecosystems, sustainably manage forests, combat desertification, and halt and reverse land degradation and halt biodiversity loss.
16. Promote peaceful and inclusive societies for sustainable development, provide access to justice, build effective accountable and inclusive institutions.
17. Strengthen the implementation of the Global Partnership for Sustainable Development.

Example 1.1

About 2,500 people live a traditional lifestyle on the Pacific island of Rotuma, as illustrated in Figure E1.1. Each morning families use two to three coconut shells in a pit oven to cook the daily taro root. Coconut shells are readily accessible around each house, they are certainly affordable, costing only a few minutes to collect and place in the sun for drying for a day or two. Coconut shells are also the byproduct of harvesting the copra for food and oil. The coconut shell fuel is reliable and definitely sustainable, but it is not modern. It has been used in the traditional society for thousands of years. The people of Rotuma would be considered 'poor' but they are not impoverished; they have food and social support, education and shelter according to their traditions. Consider that the people on Rotuma actually already have a developed society and economy.

The biggest threat to their way of life is climate change and the resulting rising sea level, which has started to kill their coconut groves. Another threat is modernization of their energy system, including diesel generators. Diesel fuel cost and availability is outside their control. The island of Rotuma is quite small and does not have room for toxic waste disposal. Thus batteries, solar PV panels, plastics and other non-degradable, toxic materials associated with modern development are likely to accumulate and degrade people's health and quality of life. As with many islands, installation of pumping engines has already caused lowering of the fresh water lens and infiltration of seawater into springs. Flush toilets that use the pumped water with septic tanks have been installed and will inevitably leak, contaminating the water supply (Krumdieck and Hamm 2009).

Is more development sustainable for the people of Rotuma? Is modernization actually a risk to traditional systems and resources? Does progress in technology reduce or increase resilience? More than 2000 years ago people lived on this island. What are the sustainable development risks for the next 2000 years?

FIGURE E1.1   The challenge of sustainable, modern energy in a traditional community as illustrated on the South Pacific island of Rotuma.

1.2.3   Engineering and Sustainable Development

The UN sustainable development goals are intended outcomes of development policy and investment. However, the business-as-usual approach to development remains largely unchanged. Countries with industrial market economies continue to consume energy and resources well beyond sustainable levels. Exploitation of resources and labour in developing countries can often lead to deteriorating conditions and more unstable governments. In this section, let us explore different perspectives on how professional engineers in all fields would contribute to the UN SDGs.

Engineers Without Borders (EWB) was founded in 2001 in Boulder, Colorado, in the United States, and since then has expanded with active groups around the world. The purpose of EWB is to work on individual projects to bring the proficiencies of engineering to developing countries. Most of the projects involve several weeks of work by students and sponsors who travel from an industrial country to a developing country. Most of the projects are addressing clean water, electricity and buildings. EWB has adopted a declaration of global responsibility for engineers to take a leadership role in addressing global challenges. The EWB model uses voluntary work and charitable donations to carry out the projects. The EWB and similar programs have been an innovation in the mindset of engineering that can result in some engineered systems being built in developing countries that otherwise would not have been possible through market forces or development policies. The EWB experience may also impress a sense of responsibility on young engineers. Given that the biggest threat to the well-being of all people is actually fossil fuel use by developed countries, innovation in mindset is needed across all engineering disciplines that leads to projects that change existing energy-intensive systems in industrial societies. The Global Association for Transition Engineering (GATE) is a new organization that is setting up student projects within corporations, organizations and city governments in developed countries to innovate, engineer and manage shift projects.

1.2.4 Challenging the Idea of Sustainable Growth

Sustainable growth is an idea that is internally inconsistent (Voss et al. 2009). Achieving a sustainable society is clearly the goal, but in the tens of thousands of years of human experience, sustainability has always meant careful management and restrictions on growth of unsustainable behaviours. There are numerous examples of unsustainable societies throughout history – and nearly all of them had a great growth period that produced grand palaces and fantastic structures, which we now visit as tourists (Diamond 2005). There are numerous examples of sustainable societies throughout history – Western societies call them indigenous, traditional or undeveloped. The fact is that sustainable societies have operational mechanisms to balance the risks and benefits of change. The most important measure of success is lack of failure. If you change what is working, it could be an opportunity to make things better or it could be a mistake and cause a failure. We are going to challenge our perceptions of the risks and benefits of the things we believe represent progress.

The historical trajectory of industrial development has led to many benefits derived from unsustainable energy and resource consumption, and unsustainable environmental impacts. Energy transition is the work of reducing consumption and restoring ecosystems. All engineering work must return a profit to a client. Who will pay for projects that down-shift existing energy use in transport, buildings and products to meet climate security restrictions? How will we quantify the benefit from not exceeding safety limits of global warming? How will communities realize the benefits from clean water, biodiversity, healthy marine or forest ecosystems? How much would it be worth to not have plastic gyres floating in the Pacific? How will our economies survive without growth?

> ## TRANSITION ENGINEERING CONCEPT 1
> *Wanting a good thing is not an effective way to stop a bad thing.*

1.2.5 Prevent What Is Preventable: The Safety Engineering Story

The best way to measure or assess sustainability is actually to evaluate the problems and risks of *unsustainability*. The logical approach for sustaining society's essential activities is to work on down-shifting the unsustainable activities. This might seem like just word play, but consider that this insight has already changed the world many times.

In 1911 a catastrophic fire at the Triangle Shirtwaist Factory in Manhattan (Cornell 2018) triggered a response by 62 industrial engineers (ASSE 2018). This disaster was not unusual for the time, when factory fires, mining disasters and machinery accidents were nearly 100 times more common than today. The Triangle Shirtwaist Factory was on the ninth floor of a building. The garment factory was filled with flammable materials and sources of ignition and only a few water buckets to put out fires. The few exits that were not locked had doors that opened inward, effectively preventing them from being opened by the panicked press of people. There was a fire escape but it did not reach the ground or another building. The fire department ladders and water pressure could only reach the sixth floor. When a fire broke out on March 25, 1911, 146 workers, mostly young women, were trapped and died over the course of several hours waiting for help at the windows, and eventually jumping to their deaths when they became engulfed in flames. Today, we clearly recognize all of these hazards as failures. But in 1911, there were no standards or safety regulations that would have prevented any of these deaths.

By October 14, the 62 members of what would become the American Society of Safety Engineers (ASSE) formed a professional organization dedicated to changing existing systems to prevent what is preventable. They developed a direct approach of identifying hazards, learning from past failures, being honest with employers and the public, and setting professional standards that superseded factory owner preferences. They all agreed to change the doors at their own factories to open outward without asking permission from the owners or managers. This and other simple safety standards improved the workplace safety so much that insurance companies quickly began requiring them in all factories. This was a turning point. If the safety standards had remained unchanged since 1911, instead of less than 200, there would be 8,600 on-the-job fatalities in the United Kingdom every year. Safety Engineering has never been about finding a safe solution; rather, it has been about identifying and changing what is unsafe. Objectively, the emergence of Safety Engineering changed the future.

Safety Engineering introduced social responsibility through professional practice, standards and self-initiated change in existing systems. This idea has been applied in other fields. Natural hazards engineering emerged after the disastrous San Francisco earthquake. Emergency management emerged after devastating hurricanes and the earthquakes in 1962 to 1972. Waste management emerged in response to toxic seepage from municipal refuse dumps. Environmental engineering to change industrial processes emerged soon after the Cuyahoga River and Lake Erie were declared dead. Reliability engineering evolved out of disasters in aerospace and nuclear industries. Energy management engineering did not exist before the 1970s energy crisis.

Figure 1.1 shows how sustainability is like safety and security, in that it is defined and measured by the number, degree and costs of system failures (Krumdieck 2013). Safety engineers work on identifying and changing products or workplaces that pose hazards. Security engineers work to manage threats to security. Sustainability engineers work on making more sustainable products and developing more renewable energy. Transition engineers work on down-shifting unsustainable energy and material production and consumption systems.

Safety	Security	Sustainability
Prevent exposure to current or immediate hazards from fire, physical environment, toxic substances, accidents and criminal behaviour.	Prevent exposure to seasonal or generational natural hazards, supply chain disruptions, logistics failures, aggression.	Prevent exposure to continuous or cumulative hazards that are too sudden for recovery or too severe for adaptation.
In homes, workplaces and public buildings. In public spaces and transportation systems.	In communities, companies, regions and states. Along trade routes, rivers and shipping channels.	In land use, biodiversity, ecosystems, built environments and natural and mineral resources.
For individuals, families and activity groups at work, at home in schools and recreation.	For tribes and governments, businesses and institutions, religious, military and service organizations.	For cultures, civilizations and nations. For living species, agriculture, forests, land and aquatic environments.

FIGURE 1.1 Safety, security and sustainability have different time, space and human scales, but issues in all three are addressed by measuring failures and making changes to reduce risks and avoid hazards, to 'prevent what is preventable'.

In 1911 workplace safety was nearly nonexistent, the economy was booming, corporations were growing in power and influence, and politicians were not responding to the protests against unsafe conditions. Factory workers had little or no voice and no protection from harm and exploitation. Today, if a person dies on the job because of a failure of safety procedures, the employer could be held liable if he or she is found negligent in following applicable standards and providing reasonable duty of care. There have also been class-action lawsuits that have held large polluters responsible and enforced payments to people who suffered health problems. But these remedies are only possible because of the safety standards.

The emergence of Safety Engineering is a useful model for understanding the pathway by which Transition Engineering will change the future. We can see the parallels between factory workers a century ago and earth citizens a century from now. The future citizens who are harmed by earlier unsustainable activities cannot file suit for damages, and they have no recourse under the law against those previous generations who caused the harm. Liability for pain and suffering is a matter of legal recourse between people alive at the same time. The sense of moral responsibility for future generations may develop to the point that regulations could require elimination of unsustainable activities. Someday court judgements will require compensatory damages be paid by fossil fuel producers to future generations. But currently there is no political power allocated to people in future generations.

Today, transition engineers are in a position to establish the standards, best practice and duty of care to address the problems of unsustainable development. The problems are global and serious. The political, economic and social context is complex and impossible to change. A century ago, engineers identified that locked doors were the reason people died in a fire, agreed to set a standard and changed the future. Today we look for energy and material down-shift projects in specific activities, and when we discover them, they will seem obvious, become standard and change the future.

Example 1.2

Most of the current safety standards and regulations were developed after catastrophic failures. For example, when children got off school buses, many were struck by passing cars. The only solution is for drivers not to hit children, which would mean they would have to stop. New technologies of flashing lights and swing-out stop signs on the buses themselves were developed to alert car drivers to the bus letting children off, and new laws were enacted to require all traffic to stop while the school bus lights are flashing, with violators receiving stiff penalties. People always valued the lives of children, but the flashing stop sign technology was not developed until *after* many children were killed, and engineer Harry Fultz became aware of the problem and came up with the safety innovation of flashing stop signs (Fultz 1958). The laws requiring motorists to stop when school bus lights are flashing and the penalties imposed for non-compliance had a marked effect on driver behaviour and greatly reduced the number of accidents. But those laws could never have been written if the technology of the flashing stop sign had not been invented.

Children's lives have thus been saved because an engineer saw a problem and thought of a technical solution, which then changed behaviour through regulation and enforcement. This example of stopping the flow of traffic to save children illustrates a central premise in Transition Engineering: as in Safety Engineering, all engineers are responsible for observing what is unsustainable and responding by applying creative problem solving and implementing changes.

1.3   UNSUSTAINABLE POLLUTION: GLOBAL WARMING AND CLIMATE CHANGE

Hydrocarbons are the basis for our energy system. Dead animals and plants were deposited in sediments between 240 and 65 million years ago during a time when the Earth's atmosphere was much higher in CO_2 and the climate was so warm that there were no ice caps or glaciers. This period of sequestration of biological carbon very gradually depleted the CO_2 in the atmosphere and contributed to the cooling climate of the planet. Even though this climate change happened over tens of millions of years, the end of the Mesozoic era had already seen the extinction of most of the land and marine dinosaur species by the time of the catastrophic event thought to be caused by a meteorite impact (Deffeyes 2001).

Extracting 100-million-year-old hydrocarbons from geologic formations and burning them for fuel leads to accumulation of carbon dioxide in the atmosphere. The concentration of CO_2 in the atmosphere is less than 1% and is measured in terms of parts per million (ppm). The Earth's atmosphere is largely transparent to the short radiation wavelengths emitted by the sun. Thus, at least 75% of the solar energy arriving at the Earth penetrates the atmosphere and is adsorbed by soil, rocks, plants and water. The exception is ice and snow, which reflect most light back through the atmosphere to space. Adsorbed light warms soils and water and this heat radiates from the Earth's surface as long wavelength infrared radiation. This is where CO_2 and other greenhouse gases (GHGs) have such a big impact. Even a tiny amount of GHG adsorbs significant infrared radiation in the atmosphere, effectively

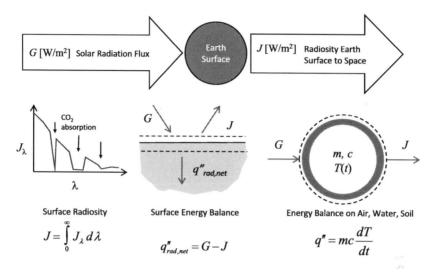

FIGURE 1.2 Heat transfer and thermodynamics of the energy balance on the atmosphere, ocean and soil of the surface of the Earth.

reducing the radiation heat transfer of the Earth's surface to space, and causing the atmosphere, land and water to increase in temperature.

Figure 1.2 illustrates the First Law of Thermodynamics applied to the planet Earth, providing a simple engineering description of the global warming problem. The radiosity, J [W/m²], of the surface is equal to the emissive power of the Earth's surface at the Earth temperature, minus the radiation adsorbed in the atmosphere at different radiation wavelengths, λ [nm]. The net radiosity is lower than the solar radiation flux, G [W/m²], resulting in radiative forcing estimated to be $q'' = 4$ W/m² for atmospheric CO_2 concentration around 350 ppm. Note that the temperature rise of the land, sea and atmosphere, dT/dt, is a function of the mass being heated, m [kg] and the specific heat, c [MJ/kgK]. The weather patterns on the Earth at any point in time are a function of the land, sea and air temperature distributions.

The climate of any place on the Earth is a function of the geography, proximity to mountains or seas, and the regional temperature distributions. The net warming energy stored due to the imbalance of incoming and outgoing radiation heat transfer rate is proportional to the change in temperature of the system. Thus, what has become a political topic is really just a heat transfer and thermodynamics reality.

Even if there were no further *growth* in fossil fuel extraction, accumulation of CO_2 in the atmosphere leading to a 2.5°C temperature rise would still result over the next 35 years. This level of temperature rise would change the Earth's climate to that similar to the Mesozoic period, the age of the dinosaurs (Hansen et al. 1981). The consensus of scientists in many fields is that exceeding 2°C in average global warming due to CO_2 concentration exceeding 450 ppm is more than 60% likely to result in 'catastrophic climate change': extinction of the Great Barrier Reef, melting of the Greenland Ice Sheet, die-off of around 50% of wild species and sea level rise and extreme weather that would threaten at least 80%

of humanmade structures. For reference, this 'carbon limit' means that 90% of known economically producible hydrocarbon must remain in the ground for the next several centuries. Thus, throughout this book, we will use 80% reduction in fossil fuel production and consumption as an engineering requirement for downshift projects to achieve the energy transition.

If the Earth had no atmosphere, then the equilibrium surface temperature would be about $-20°C$. Lucky for us, about 150 W/m^2 of the solar radiation adsorbed by the land and sea is adsorbed by the atmosphere, forming an insulating layer that keeps the average temperature of the Earth at 15°C, where it has been throughout the past 8000-year history of human civilization. The most important GHG in relation to energy systems is the CO_2 produced by burning the fossil carbon fuels. Fossil carbon that is taken from geologic deposits and burned for energy *accumulates* in the air and water, changing thermal properties of the atmosphere and the chemical composition of the ocean. A billion metric tons of solid fossil carbon is referred to as a giga-ton (Gt-C), and when burned produces 3.67 Gt-CO_2.

From 1750 to 2015, the accumulated GHGs, which include CO_2, methane, nitrous oxide and humanmade refrigerants and other compounds, have increased the rate of heat stored in the Earth system by $dq''_{rad, net} = 2.8$ W/m^2, and the global average temperature has increased by 0.8°C. This seemingly small change in average planetary energy balance has already increased the occurrence of extreme weather events like floods and droughts, heatwaves and hurricanes, and has accelerated the melting of glaciers and polar ice.

The CO_2 additions to the atmosphere from burning fossil fuels and from the manufacture of cement have been at least 10 Gt-C/yr for the past decade. Clearance of forests, especially by burning, and soil disturbance for agriculture also generate about 1.9–2.2 Gt-C/yr. The global plant uptake of CO_2 averages 4 ± 1 Gt-C/yr depending on weather conditions. The ocean is estimated to be adsorbing 2.0 ± 0.8 Gt-C/yr, which is mitigating global warming, but this process is causing acidification effects that are harming coral reefs and other ecosystems. Some forest areas that were previously cleared are being allowed to regenerate and are estimated to be absorbing 0.5 ± 0.05 Gt-C/yr (ESRL 2017). The imbalance between humanmade emissions and plant uptake is around 5 Gt-C/yr, and this imbalance, accumulated over the past 70 years, has caused the current CO_2 atmospheric concentration that has not existed for over 1 million years (NASA 2015).

The US National Oceanic and Atmospheric Administration (NOAA 2017) data shows that 2015 CO_2 emissions from fossil fuel and forest burning were the highest on record, at over 12 Gt-C/yr, an increase of 3% from the previous year. The resulting global warming is rapidly occurring, with 2016 representing the third consecutive year in a row with record global temperatures, with average global temperature 1.1°C above the late nineteenth-century average. The effects of humanmade emissions have long been well understood to pose a risk to food production, coastal cities and human health, as demonstrated in Table 1.1. Note that the safe range of accumulated CO_2 in the atmosphere of 350 ppm was surpassed before 1990. As a result, the extreme weather events, unprecedented storms, record high temperatures, floods, droughts and wildfires are now generally understood by climate scientists to be a result of added thermal energy in the oceans, land masses and atmosphere.

TABLE 1.1

An Abbreviated History of Climate Science, the Annual Extraction of Fossil Carbon and the Atmospheric Concentration of the Carbon Dioxide Greenhouse Gas

Date	Parties	Statement	Fossil C (Gt-C/yr)	CO_2 (ppm)
1896	Svante Arrhenius, Swedish scientist	Global warming is possible due to CO_2 released from burning coal in industry	0.10	280
1924	A. Lotka, physicist	Calculates that burning coal at current rates will double the atmospheric CO_2 within 500 years	0.30	286
1948	Scientific debate	Earth and oceans are 'self-regulating' and natural balance may absorb anthropogenic CO_2 emissions	0.80	300
1958	C.D. Keeling	Measurement of CO_2 at Mauna Loa, Hawaii	1.20	315
1975	Manabe and Wetherald, NOAA	First three-dimensional model of Earth climate including effects of GHGs	2.40	330
1985	UN Conference ICSU	Global warming is inevitable due to existing CO_2 accumulation	3.60	346
1990	49 Nobel Prize winners	'… the build-up of various gases introduced by human activity has the potential to produce dramatic changes in climate …'	4.40	354
1990	IPCC: 170 scientists	60% reduction in CO_2 emissions needed to stop temperature increase	4.40	354
1995	IPCC consensus of climate scientists	There is a discernible human influence on global climate (*hottest year on record*)	5.40	361
1997	Kyoto Protocol	International agreement to limit emissions to 1990 levels by 2012 (*1998 hottest year on record*)	5.86	364
2001	IPCC third Assessment Report	Most of the warming over the last 50 years is attributable to human activities	6.90	370
2001	US National Academy of Sciences	Temperatures are in fact rising due to human activities	6.90	371
2001	US Global Change Research Program	Higher probability of extreme weather, drought, flood, storms and sea level rise is now occurring	6.90	371
2002	Europe, Canada, Japan	Ratify Kyoto agreement, Europe sets up carbon trading, Japan reduces CO_2 emissions	6.97	373
2004	Numerous scientists	Arctic sea ice has thinned by 3.1–1.8 m, and extent of summer ice shelf has shrunk by 5%	7.76	378
2012	NOAA	Hottest summer on record and most severe drought over largest area of the United States	9.61	394
2012	Dr. James Hansen, Chief Climate Scientist, NASA	'The target that has been talked about in international negotiations for two degrees of warming is actually a prescription for long-term disaster.'	9.61	394

(*Continued*)

TABLE 1.1 (*Continued*)
An Abbreviated History of Climate Science, the Annual Extraction of Fossil Carbon and the Atmospheric Concentration of the Carbon Dioxide Greenhouse Gas

Date	Parties	Statement	Fossil C (Gt-C/yr)	CO_2 (ppm)
2014	IPCC fifth Assessment Report	85% of known fossil fuel reserves must remain un-burned in order to limit warming to 2°C	9.82	400
2015	UN COP21	Paris Agreement of 195 nations to limit global warming to well below 2°C by rapidly reducing emissions	9.83	403
2016	NOAA, NASA	Third consecutive year of record global temperatures 1.1°C above late nineteenth-century average	>10	405
2017	NOAA	Hottest year on record without an El Niño event; 40% increase thermal forcing since 1990	>10	408
2018	NOAA	April monthly CO_2 exceeded 410 ppm; Arctic sea ice 40% lower than first measure 1977	>11	>410

The biggest challenge for energy transition is generating the innovation that is needed to achieve dramatic reduction of fossil fuel production to stall the accumulation of CO_2 and other humanmade GHGs well below 450 ppm (Hansen 2009). If 800 Gt-C in total are extracted and burned, the result will be global warming of 2°C, and this means a 66% probability that polar ice and glaciers will melt enough to raise sea level above where coastal cities can cope. This level of CO_2 pollution would also cause numerous other deleterious changes, foremost among them the acidification of the oceans and loss of most coral reefs. Because even 1.5°C of warming could cause tundra to melt, in turn causing additional methane emissions, the probability of initiating catastrophic climate change will be at least one in five if 565 more Gt-C from fossil carbon sources are added to the atmosphere after 2015.

Fossil fuel combustion must be reduced by 80% by 2050 in order to have a chance of avoiding run-away climate change (IPCC 2013). The climate science shows that CO_2 concentration below 350 ppm would reduce the risks of damaging climate change, and this level has been accepted as a 'safe target'. Like the safety limit concept used in engineering, the 350 ppm level of CO_2 in the atmosphere has a statistically relevant possibility of not causing a catastrophic failure. The scientists that study the climate system dynamics describe a system much like a mechanical system that has an increasingly higher probability of a catastrophic failure the closer you push the system out of the stability region.

TRANSITION ENGINEERING CONCEPT 2

Global average temperature rise of 2°C is not a target; it is a failure limit.

The most straightforward way to approach this important work is to focus on the solid carbon production limits for climate stability put forward by scientists (IPCC 2013). Recent estimates for the known economically recoverable reserves of coal, oil and natural gas are estimated at 2,800 Gt-C. This represents the fossil carbon that producers are planning to bring to the market. Higher concentration of GHGs in the atmosphere means more solar energy in the planetary atmospheric and ocean systems. This means changes in climate as average temperatures rise, and more severe storms as weather systems are driven by more thermal energy.

Extreme weather events have already brought about unprecedented hurricane, flood and fire events around the world. However, a particularly salient risk is the melting of the Greenland Ice Sheet and the resulting sea level rise. The Greenland Ice Sheet is 2.3 km thick and contains around 8% of the fresh water on the planet. Much of the polar ice is frozen seawater, and while the extent of the polar ice has been shrinking, sea ice melting does not contribute to sea level rise. However, the Greenland ice is fresh water that will flow into the sea and increase the volume of oceanic water. By unambiguous analysis, if or when the Greenland Ice Sheet melts, sea level will rise by 7 m. Considering that at least one-third of the world's population lives in coastal zones, the human impact alone represents an unprecedented global catastrophe.

For the past decade, every study by scientists from different countries and with different specialties, using a variety of methods, have all concluded that Greenland ice was stable until 1990 and since then has been melting at an accelerating rate (Luthcke et al. 2006). The amount of global warming to date will likely result in around 13 cm of sea level rise over the next 30 years, even if fossil fuel use were curtailed completely today. The current international agreements under the Kyoto Protocol, the Copenhagen climate conference and the COP21 represent non-binding pledges by political leaders of over 150 nations. Even if achieved, the agreed actions would have a 60% chance of resulting in 4°C global warming. In summary, the requirements for the safety of all of humanity's infrastructure and the well-being and the sustainability of natural ecosystems and species *requires* 80% reduction of fossil fuel production across the board within two decades. This is the energy transition.

1.3.1 UNDERSTANDING THE 2°C FAILURE LIMIT

Let's use a thought experiment for the purpose of understanding the problem of climate change and fossil fuel extraction. Imagine that we lump all of the benefits to humanity being provided by the planetary troposphere together into one symbolic and very useful structure – a bridge. Figure 1.3 illustrates the cumulative loading problem for a structure.

A bridge can be used by all people and provides social benefits of mobility, access to goods from across the river, economic trade relationships and even perhaps a good place for fishing. Most people just use the bridge and do not give a thought to how it was made, how it works or what the risks to its structure might be. From an engineering perspective, the bridge is a stable and permanent part of the transport infrastructure and will continue to perform as long as it is maintained and as long as the safe load limit is not exceeded. The civil engineers who designed and built the structure

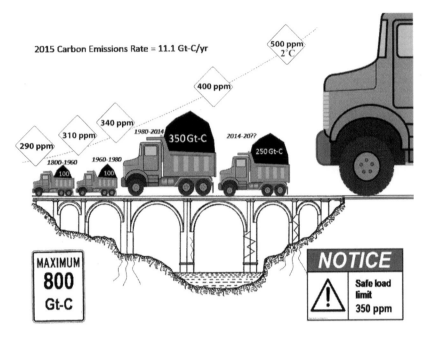

FIGURE 1.3 The bridge represents the global climate and ecosystems, the trucks are the fossil carbon extracted from geological features and combusted in the timeframes shown, the caution signs are the measured atmospheric CO_2 concentration, and the really big truck trying to get onto the bridge represents the 2,800 Gt of fossil carbon known to be economically producible. Note that the trucks cannot get off the bridge, and that the loading to date has exceeded the safe loading limit and has caused stresses and fractures in support structures and foundations.

would not be able to predict exactly how the bridge would fail if it were overloaded with coal trucks, but they would be very clear that you should not put enough big trucks on the bridge that the safe load limit is exceeded.

If the safety limit is exceeded, the structure might not fail, but the probability of failure somewhere in the system is above an acceptable level. The maximum loading to failure is another familiar concept in engineering. There is only a small probability that a system loaded to the failure limit will continue to be safe for normal use. At the current global fossil fuel extraction rate of over 10 GTC/yr, the climate failure limit will be reached before 2040. The really big truck heading for the bridge represents the economically producible fossil fuel reserves. The energy transition is clearly a project of down-shifting fossil fuel use faster than any political leaders are anticipating.

1.4   OIL SUPPLY AND PEAK OIL

Oil is the linchpin fuel for the current global economy. The world oil production capacity increased apace with demand for liquid fuels most of the last 100 years but reached a plateau point from around 2006. Think about the finite nature of oil and the fact that more than half of the oil in easily accessed reservoirs has been used. Consider this statement because it seems logical then that 100 years from now, the age of oil will be over. Over the past century oil production has grown exponentially, and over the next century production will decline to a very small fraction of current volumes.

Prominent petroleum geologists like industry experts Campbell and Laherrare (1998), Princeton University professor Deffeyes (2001) and Uppsala University professors Aleklett and Colin (2003) have published analyses showing conventional oil production peaking in the first decade of this century. Experts using different assumptions and data have reported a range of peak production dates and subsequent decline rates. Of course, there is uncertainty about future supply scenarios, but there is total agreement that the available oil supply within the lifetime of all road and airport infrastructure will be significantly less than today (Chapman 2014).

A method for dealing with a multitude of experts is to include all of the information in a meta-analysis (Krumdieck et al. 2010). The method assumes that all of the experts have a reasonable accuracy based on their assumptions and data, and that every expert analysis is valuable. A probability distribution function, like a Rayleigh distribution, can be fitted to all of the expert reports and a probability space generated. Figure 1.4 shows the probability space for oil supply according to a recent survey of published expert analyses. The lowest curve is the oil supply that 97% of experts agree will be available. The top curve represents production volumes only 3% of experts believe are possible in future years. Essentially, oil production rates in future years are not considered possible by *any* experts above the top curve. This type of probability analysis can be used for other finite resources. Decision-makers can also use the probability analysis because they can

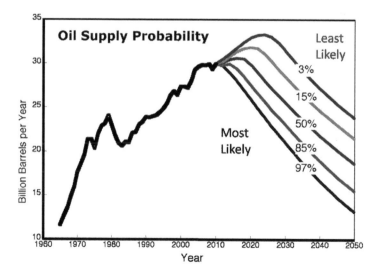

FIGURE 1.4 Probability of future oil supply generated from meta-analysis of oil expert analysis. (From Krumdieck, S., et al., *Transport Res. Part A*, 44, 306–322, 2010.)

decide their risk position and then use the associated future oil supply level over the planning horizon of interest.

Oil supply security has been a serious issue for most countries for more than a century, caused as much by political and economic instability as by resource availability. A short-term supply disruption or price spike is called an oil crisis. The long-term depletion of oil reservoirs is not in question, but in the near term there is a period called peak oil, when the oil production rate stops increasing and begins to decline. The peak of the oil production does not involve scarcity, as the production volume is still near the historical high. However, the divergence between the production rate and demand growth can cause limited spare capacity in the world oil market, and this situation results in rapid escalation of the price. Of course, a price increase is good for oil companies and oil-producing countries, as their revenues rise without making any more investment in exploration or developing oil fields. However, an oil price rise is bad for consumers and businesses because they are suddenly spending more to do exactly the same activity and to generate the same returns. Such a large unexpected cost, which consumers have no way to avoid, causes a recession in the economy, as spending slows down in other areas.

1.4.1   THE OPEC OIL EMBARGO AND THE 1970S ENERGY CRISIS

The production of oil in the United States has been controlled at affordable prices by various organizations since 1910. The decisions made 60 years ago about building roads, dismantling urban trams, letting railroads decline, and allowing low-density land use reflect the perception that low-cost oil would last forever. The average price of oil remained below $3.00 per barrel, largely because of government price controls, until 1970. On 5 October 1973 the Yom Kippur War started when Syria and Egypt attacked Israel. The United States and most western countries showed support for Israel. Several Arab nations, members of the Organization of Petroleum Exporting Countries (OPEC) plus Iran imposed an oil embargo on western nations, and curtailed their production by 5 million barrels per day (mbpd), or about 10% of total world supply.

There had been no spare production in the United States, the largest oil-producing nation, since 1971, and other non-OPEC producers were able to increase production by only 1 mbpd, resulting in an overall 7% decline in world oil supply. By the end of 1974, the price of oil was over $12 per barrel. OPEC did not increase production again after the crisis, but there was a rapid increase in world supply from other areas like the North Sea, Mexico and Alaska. Despite the increased supply, the price remained in the new price range of $12.50–$14.50 per barrel. The Iranian revolution followed shortly after the invasion by Iraq in September 1980, which led to a decrease in production of 6.5 mbpd from the two warring nations, representing a further 10% decline in world supply from 1979. This led to a second oil supply shortage, with long lines at gas stations and shortages around the world. By 1981, crude oil prices had more than doubled, to $35 per barrel, and the global economic recession of the early 1980s was the most severe economic crisis since the Great Depression of the 1930s.

In the United States, this series of events, which came to be known as *the energy crisis*, did not apply to transportation only. Electricity demand had grown rapidly since World War II, with a building boom and an explosion of new energy uses for new conveniences like air conditioning, refrigeration, washing machines and other appliances. Most of the manufacturing of the new array of consumer goods was also done in the United States, adding to rapid electricity and fuel consumption growth. Significant generation capacity had been added using nuclear power plants. In the decade just before the energy crisis, diesel power plants had been built to deal with peaks in air conditioning loads. With no spare oil production capacity in the United States, the OPEC oil embargo also caused electricity price increases and shortages of as much as 20% in some places.

The energy crisis generated great interest in alternative energy sources and resulted in new standards and regulations on energy efficiency for buildings, appliances and vehicles. The US government spending on energy research and development (R&D) went from $5.6 million in 1974 to $715.3 million in 1980 and the Solar Energy Research Institute was established (US IEA 2017(a)). There was a general sense that, with a 'man on the moon' type of collective effort, a technology revolution in alternative energy would be inevitable. The energy crisis also started a political obsession with energy independence as a national security objective that persists to this day. The realization that all critical activities rely on finite fossil fuels has influenced thinking about the future implications of depletion. Energy engineering emerged as a new field, primarily within mechanical engineering, with universities offering courses on energy conversion, air pollution, alternative energy and energy management. However, as oil fields in the North Sea and in non-OPEC nations were rapidly brought into production, the world oil supply increased again, and the oil price declined to a new price range below $30 per barrel. Throughout the 1990s, it became politically unpopular to worry about energy shortages. In fact, the discussion around fossil fuel use and the impact of development of the resources in remote and higher risk environments became polarized and was framed in terms of the environment versus the economy.

1.4.2   OIL AND THE ECONOMY

There have been 11 economic recessions since World War II. One of them followed the bursting of the dot-com investment bubble, and an oil price spike preceded the rest (Rubin and Buchanan 2008). Oil production from Mexico, the North Sea and Alaska went into decline in 2006, Middle East production did not increase, and increased production from Brazil and Eurasia could not fill the gap between crude oil production and refinery futures requirements. Oil refineries are the real customers for crude oil; the refining processes cannot easily be changed, so refineries are willing to pay significantly higher prices to secure contracts for oil deliveries in future months rather than shut down. By 2008, the oil futures markets were reacting to the global spare capacity margin going below 1 mbpd and the price went over $100 per barrel for the first time in history (US EIA 2015). The global economic crisis, which started in 2008, was a much more severe economic downturn than the 1980's recession. The oil price remained over $100 per barrel for four years until 2014. With these high oil prices presumably remaining stable, oil companies invested in developing unconventional reserves that had previously been too expensive to produce and bring to the market. Unconventional hydrocarbon production includes oil fields in deep water off shore; the Athabasca tar sands; and low grade, 'tight' oil that is actually not in reservoirs but locked in source rock, requiring hydraulic fracturing.

Demand for oil declined in most countries from 2008 to 2014, surprising most economic analysts. In the United States, consumption in 2014 was lower than in 1997, and this would have been completely unthinkable prior to 2006, when consumption of conventional oil peaked. In 2003, the US Energy Information Administration (EIA) predicted that US oil demand in 2025 would be around 30 mbpd, but the 2015 Annual Energy Outlook predicted nearly stagnant demand, with 34% lower projection of future demand. The idea of a peak and decline in oil supply was seen as an economic doomsday scenario before the 2010s. But some economists now have a positive outlook regarding the $250 billion *not* spent on oil in 2025 (Cox 2015).

Most neoliberal economists do not recognize a link between oil and the economy (King 2017). However, it is not hard to apply some simple logic to reveal how a sudden jump in oil prices can trigger a recession. All companies, organizations and households use oil, but there are only certain industries, like trucking and aviation, where fuel costs are a substantial cost of doing business. Producers and service providers simply pass on the costs of doing business to customers; so how does an oil price spike cause an economic recession? There are two main factors: the cumulative behaviour of all actors and the specific availability of capital. All organizations develop annual budgets, based on the previous year's incomes and expenditures, which have built-in surplus at the end of the year to use for maintenance or expansion. These expectations for investing a bit of surplus back into the company need to be at least 3% of the company gross receipts in order to maintain the capability to do business. For example, a restaurant needs to replace broken glasses, repair an old refrigerator or freshen up decorations. A manufacturing company may have plans to expand by purchasing a new milling machine and hiring a new operator. An unanticipated new cost, like the 2007 and 2008 doubling of costs for fuel, would not cause bankruptcy in a company or layoffs at a university, but they would eat up the surplus for the year. The result would be decisions *not* to buy new equipment or hire new staff in each individual business, and thus throughout the entire economy – and that is a recession.

Capital is the term used for money, usually borrowed from banks. Capital is used to build new buildings or factories, purchase new trucks to expand a freight operation, or purchase ovens and cabinets to start a new bakery business. When the price of oil doubled, the United States was consuming about 20 mbpd. The oil price rise of $50 per barrel in 2008 meant that $380 billion would have gone to operating costs for exactly the same revenue-generating activities. Oil companies and oil-producing countries tend to save profits rather than lending capital or investing in other enterprises like factories or bakeries. The net result of the oil price spike was an overall reduction of global capital available and a contraction of business activities. The global economic crisis also involved the collapse of financial schemes that were dubious and illegal, but the role of the oil price spike in knocking over the house of cards is not hard to see. Energy transition innovations that help companies and organizations deal with oil price spikes could benefit the companies, as well as the whole economy. Such innovative shift projects would involve energy audits of oil use and development of adaptive oil use management plans as well as reducing the amount of oil used in the operations.

Example 1.3

Let's explore the relationships between oil price, recession and energy shift. A bread bakery is located in a central business district and has a good reputation for hand-made loaves from organic ingredients. The bakery oven capacity is 200 loaves per day. The bakery currently has average sales of 100 loaves per day to people who drive to the shop. The business manager is looking for ways to increase sales volumes to better utilize the production capacity. The manager expands the market to full capacity by purchasing a van, hiring a driver and delivering loaves to cafes and restaurants around the city. The profit margin (net profit/total revenues) is 5%, and the fuel is 6% of total operating costs.

Then – the price of fuel doubles. The cost of fuel increases to 8% of operating costs, and the profit margin drops by 1%, which is not good. The manager now has to think about asking staff to take a wage cut, or reducing the quality of the ingredients, which could damage the brand and reduce sales. A switch to an electric delivery van would reduce the transport costs back to the level before the oil price rise, but the electric van is expensive, and there are no banks that will make a loan to a small business just to carry out the same level of operations. Capital is invested in order to extract value from growth.

In response to the oil price spike, the city council is building new cycle ways. The manager supports the effort and lobbies to get the cycle way to the bakery storefront. The manager sells the van and builds a cycle parking facility and outdoor seating with the proceeds. The increased cycle traffic, plus drive-up traffic, results in sales of 150 loaves per day, and the profit margin increases to 21%.

1.5 DISCUSSION

The *mega-problems* of risk for all of the world's energy systems are global warming and oil supply. The mechanisms of global warming and the consequences of climate change have been well known for many decades. In fact, more than 30 years ago the nations of the world agreed to take urgent action and curtail emissions. The global scientific community, world leaders and the public may agree that emissions must be dramatically reduced, but fossil fuel exploration and production are still heavily subsidized. The countries that have fossil fuel production operations do not have plans for curtailing them. Over the past decade, the issue of oil extraction peaking and declining became a topic of discussion and concern. In fact, it has become apparent that conventional crude oil production peaked in 2006–2008 and is declining. Unconventional liquid hydrocarbon resources that were too costly and environmentally damaging to be considered 10 years ago are now being developed. Industrial society requires vast quantities of fossil fuels. However, curtailing production of fossil fuels is the only way to mitigate the climate risks facing all societies.

1.5.1 THE MONKEY TRAP

Bush hunters know what fruit monkeys love the most. The hunters tie a jar with a small neck to a tree and put a large piece of the favourite fruit into a jar. The monkey can smell the fruit, he can squeeze his hand in and grab the fruit. But he can't get his hand out when he has hold of the fruit. The monkey will not let go of its favourite fruit.

The hunter approaches with his club. The monkey knows that the hunter is dangerous. First the monkey stays very still hoping the hunter will not see him. Then he screeches and shows his teeth and protests. The monkey jumps up and down and pulls against the rope holding the jar. As the hunter gets closer, the monkey could let go and run away, but instead he yells louder and gets more panicked. By the time the hunter is close enough to use the club, the monkey is so distressed he can't see that he must let go of what he likes most in life in order to live another day.

We can understand what it is like for the monkey. At first, there is no indication of danger. The food in the jar is the best food. It is tasty and free and you do not have to go hunting around the jungle for it. So why wouldn't you reach in and grab it! Then you get surprised that you can't seem to get the food out of the jar. At first you are not alarmed, just really curious about how this works and how to get the food out. And then the hunter arrives and you are distressed and distracted by the immediate threat. Your focus on the deadly risk becomes so intense that you totally lose sight of the fact that you have actually trapped yourself. You can save yourself, but you have to let go of what you want most, and what you thought you were going to have. You need food to survive, but you also need to let it go to survive. Your future depends on what you do right now. There is one future where you let go of your favourite food and run to safety. There is another future where you don't.

There is no question that fossil fuels are the best energy source and always will be. There is no question that using fossil fuels lets us have lifestyles more comfortable and convenient than most of the kings in history. There is no question that we need fossil fuels to keep our lives going in the ways we expect. And there is no question that the risks are real and present. We know that if we down-shift the production of fossil fuels dramatically it would mitigate the risks. There is one future where fossil fuel use is curtailed for consumption and used only as an essential work fuel. There is another future where we don't let go of doing what we want until it is too late.

The next chapter describes even more risks of unsustainability, but then the rest of the text answers the question, 'How do we achieve the transition beyond fossil fuel?' One important thing to keep in mind is that there are no *solutions* to a monkey trap. A monkey trap is not a *problem*; it is not a thing that happened to the monkey. A monkey trap is a *situation* that resulted from a series of choices and a set of realities. Transition Engineering is best described as strategic change management. The solution already exists – curtail fossil fuel production. Transition Engineering is the work of discovering *how*.

2 Problems of Unsustainability

Define the system and relationships to other systems.

By definition, any trend that cannot continue is unsustainable and will change. The historical growth of energy use has driven unsustainable growth in resource use, land use, and environmental impacts. Energy system investments made today will face interconnected environmental, social and economic issues throughout the useful life of those energy systems. Although this book does not directly deal with these other issues, it is important to keep them in mind. Energy is a politically charged engineering field. As transition engineers develop down-shift projects in existing energy systems, they will face challenges in communicating the different types of costs and benefits.

This chapter introduces the historical treatment of sustainability as a concept and identifies some of the trends and issues that are motivating the energy transition. The issues of unsustainable development are wide-ranging and we will look at questions of population, water, food, forests and other unsustainable trends. The policy framework and business environment of energy transition are still emerging, but engineers have immediate responsibilities for developing the down-shift projects for energy production and end-use technologies that collectively achieve the energy transition. The energy transition will require more innovation, resourcefulness, creativity and courage as well as more renewable energy.

2.1 REVIEW OF SUSTAINABILITY PRINCIPLES

The energy choices made in the near future are among the most important of any choices in human history. While energy development questions may appear to be mostly technical and economic, there are broader considerations. Sustainability considerations reflect priorities in our society as well as our attitude toward future generations. A sustainable energy future appears feasible only if we develop an overall social and political strategy that includes renewable energy development, combined with energy conservation and adaptation of our affluent lifestyles to greatly reduced energy consumption (Rojey 2009).

2.1.1 MANAGEMENT OF COMMON RESOURCES

Allocation of limited common resources is an ancient practice that is essential for sustainability. There are villages around the world where the same families have farmed the same land and lived in the same homes for over 600 years. An agricultural township would have shared grazing areas, forest resources, walkways, bridges and access to water. Management of these common resources ensured that users did not overtax resources. Where resources have become scarce, protection and allocation systems have been developed and ensconced in cultural norms. These basic principles have worked well for thousands of years, and where they have broken down, the society has faced crisis. Traditional systems that control individual access to shared resources can be broken down by the the emergence of new technologies that allow faster extraction or by the emergence of new economic relationships that provide incentives for increasing extraction. If extraction continues to the point where there is no more extraction possible, the resource becomes constrained.

Water is an example in modern times where constraints have been reached. For example, in the western United States and Australia, the demand for water has increased beyond supply. Systems have been developed to allocate water to end-users, educate users about conservation and charge for water use. Engineering research led to new low water-using technologies and appliances, which are now mandatory in many places. Another emerging example is the number or river restoration projects in Europe, that are limiting private land uses and discharges in order to improve water quality and aquatic and bird life all along rivers (Pearce 2013). These projects on rivers in Sweden, the United Kingdom, across Germany to France and the Black Sea have required transition innovations in the practices of civil engineering.

For many years, engineers built dams and levees to try to control floods, drained wetlands to increase land use, and dredged and straightened river channels to increase navigation. Today, the engineering to restore these long riparian corridors, while managing local private interests, involves innovations in modelling as well as community and business engagement and policy development. These 'restoration engineering' practices would never have been anticipated in the last century.

2.1.2 ENVIRONMENTAL REGULATIONS

Regulations to limit environmental impacts and protect public health are key tools for changing unsustainable activities. Punishment of people who pollute rivers and lakes is an ancient practice around the world. One of the first environmental regulations in the United States was the 1899 River and Harbor Act banning disposal of things like dead animals in navigable waterways. As industries grew up around the country, states, local counties and cities have worked with health boards to monitor and regulate pollution discharged into rivers.

Example 2.1

Depletion of atmospheric ozone was identified as the result of the use of hydrochlorofluorocarbon (HCFC) refrigerants and propellants. HCFCs were actually a great invention dramatically reducing the explosion hazards and toxic exposures of the previous refrigerant, ammonia. By the 1970s, the scientific evidence was clear that HCFCs were depleting the earth's ozone layer that protected the biosphere from ultraviolet (UV) radiation.

There was great resistance from numerous industries and government to limiting production. When ultraviolet B (UVB) exposure was linked to skin cancer, and then the ozone hole was discovered over Antarctica, the issue of HCFCs became nightly news. The issue largely focused on spray cans and became polarized between those who predicted global catastrophe and those who said there was no evidence that these harmless chemicals could be changing the Earth's atmosphere. Substitute propellants were developed and brought to the market, and public preference for deodorants and hair sprays that were 'ozone friendly' meant in turn that the environmental and health risks began to outweigh the costs to industry of making the change in refrigerator technology. In 1987, the Montreal Protocol on Substances that Deplete the Ozone

Layer was ratified by most nations around the world and required corporations to phase out production and use of HCFCs by 2030 (Morrisette 1989).
The ozone hole has stopped growing in recent years, but people who live in
South America and New Zealand now live with dangerous UV exposure and
increased skin cancer rates.

The 1970s saw several pieces of US federal environmental legislation, including the
Clean Air Act (1972) and the Safe Drinking Water Act (1974), which led to marked
improvements in air and water quality, respectively. The US Clean Air Act sets minimum levels of air quality that local areas must achieve. Most countries require consent
for industries to discharge pollutants and have some degree of air and water pollution
monitoring. Nearly all of these regulations do not prohibit pollution by industry; the
main purpose is to set the level of allowable discharges and require reporting of emissions. Research into the overall economic impact of the Clean Air Act consistently
shows that job losses were not due to the regulations, new jobs were created because
of improving technology and the public health benefits between 1990 and 2010 were
between 30 and 300 times greater than the cost to industry (Greenstone 2002).

The most important lesson to learn from history is that environmental regulation is usually developed well *after* the risk is known and after technically feasible
alternatives have been developed. This pattern has many examples: the pesticide
DDT, tetraethyl-lead additive in gasoline, particulate and sulphur dioxide emissions
for coal-fired power plants, and volatile hydrocarbons and nitrous oxide pollution
from automobiles. The companies that were innovators of transition technologies
(e.g., electrostatic precipitators or catalytic converters) and transitioned their systems
ahead of the regulation were in a superior business position to those who fought the
changes. Thus, one of the guiding principles of Transition Engineering is to seek out
scientific evidence and work on changes to existing technologies and processes to
mitigate the risks and take a leadership role in developing and enacting regulations
and standards. If you want an example, research the history of the catalytic converter
and the role of Ford Motor Company.

2.1.3 Sustainable Business Practices

Triple bottom line (TBL) accounting is an idea that was developed in the late 1980s
and 1990s as an economic approach to sustainability. The TBL practice accounts for
the financial, social and environmental benefits or costs of corporate operations for
all stakeholders. Stakeholders are workers, customers and people who live near factories or mines. The financial performance of the corporation is obviously the main
bottom line as all corporations are operated to return profit to shareholders. Social
and environmental bottom lines are not as easy to quantify, so corporations normally
set out principles for their practices so that they do not exploit resources or people in
a way that negatively affects well-being.

Some companies monitor the environmental and labour practices of their suppliers to make sure fair wages are paid, child labour is not exploited and environmental

regulations are not violated. Fair trade certification has been developed to ensure that suppliers in less developed countries receive a fair price for their goods and labour. Ecological footprint, carbon footprint, lifecycle assessment (LCA) and cradle-to-grave analysis of goods have been developed to quantify the energy and material use and emissions associated with products (USDOE 2014). These types of accounting approaches have yielded real improvements in sustainability in instances where the companies set quantifiable sustainability goals. Particular successes, like 3M's energy savings, are achieved by rewarding the company's engineering staff for carrying out audits; proposing action plans; and making changes to supply chains, materials, designs, processes and operations.

Social businesses and B-corporations are for-profit companies that have social and environmental well-being as part of their legally defined goals. Many corporations have sustainability managers or sustainability champions charged with carrying out TBL accounting and reporting on company and supply chain footprints. Can you find any regulations requiring corporations to consider sustainability or to ensure the well-being of stakeholders?

Consideration of sustainability is rapidly increasing in business and industry and is even seen as a necessity to attract and retain top talent and to foster a climate of heretical innovation and creative problem solving (Winston 2014). In 2014, Winston reported that 56 of the world's 200 largest corporations had set clear goals for carbon reduction in their manufacturing, product use or supply chains, representing about 3% annual carbon reduction. A growing number of case studies of businesses making sustainability changes and increasing energy efficiency and reducing waste can be found. The demand has never been higher for engineers who can carry out energy audits to analyze manufacturing processes and supply chains; identify inefficiencies; optimize technology and operations; and save money in energy, water, materials and disposal costs (Allen and Shonnard 2012).

2.1.3.1 Business Transition Strategies

Winston (2014, p. 1683) outlines business strategies for companies executing big, science-based goals needed for transition projects:

- Understand the science and the global goals when they are clear.
- Assume the scientific recommendations will become less flexible.
- Collaborate with others when the science isn't crystal clear (but is indicative).
- Understand where context matters.
- Set the goals from the top of the organization.
- Understand the competition, whether you have an opportunity to get ahead or if you are falling behind.
- Generate a lot of ideas.
- Go for big scale and quick payback rather than incremental savings.
- Use available tools to create your targets.
- Believe that massive, radical reductions are not only possible but also profitable.

2.1.4 POLICY AND POLITICS OF SUSTAINABILITY

Policy can refer to the established goals of a private company or public organization. Politics is largely the process of deciding how to spend public money. In sustainability, both private and public policy tend to fall into the areas of green buildings, clean energy, transportation, climate protection, water systems, foodservice, waste minimization and recycling, operations, and environmental purchasing. Public concern about environmental issues is expressed through citizen's initiatives; environmental protection groups; organized protests and demonstrations; and, more recently, the social media. Local environmental and social activist groups applying resistance to specific fossil fuel extraction operations is often referred to as NIMBY (not in my back yard). There have been some visible protests around fracking, coal mining and offshore drilling in many countries (Princen 2015). Citizen actions against pollution, environmental destruction and climate change have been occurring for decades. There have been some successes of specific pollution regulations and restrictions on mining or drilling in a few environmentally sensitive areas. However, the evidence is clear that global accumulation of greenhouse gases (GHGs) has not been arrested though citizen actions or organized calls for political action. At the time of publication, there have been no protests or calls for action aimed at the engineering professional organizations. Imagine a protest on the scale of the global 15 March 2019 Schools Strike 4 Climate being aimed at the Mechanical, Civil, Chemical and Electrical Engineering societies in countries around the world demanding that the people who are designing and operating the systems that produce and consume fossil fuel change their practice and take responsibility for achieving rapid energy transition. That would be a surprising change of perspective.

The number of research papers with the term *energy transition* in the title has exploded to over 12,000 in 2014; notably most of these papers are in economics, policy and social science journals (Araujo 2014). Many governments in Europe and large industrial nations including Japan and China have set carbon emissions reduction goals. A growing number of cities have declared targets for carbon neutrality. Policy designers in all countries understand that the next International Panel on Climate Change (IPCC) report will not relax the current Fifth Assessment urgent call for 80% reduction in GHG emissions by 2050. Most notably the United States has provided much of the science around global warming and other pollution effects and took on a leadership role during the 1990s, but it has recently experienced a collapse in energy transition policy.

Economists have great influence in political thinking. An imposed price on carbon has long been the one policy instrument thought to be effective for achieving carbon emission reduction through market forces. So far, there have been two approaches to establishing a cost for pollution: carbon tax on fossil fuels and emissions trading schemes. The carbon tax would make cheap fossil fuel more expensive, in the hope that renewable energy would be more competitive. The carbon tax would be straightforward and could provide a revenue stream for projects to improve efficiency or increase renewable conversion. But given the past history of demand reduction and economic recession during periods of increased oil, coal and gas prices, the carbon tax is not popular among politicians.

The European Emissions Trading Scheme (ETS) issues carbon emission credits to businesses, then sets up a market for trading the credits. The theory is that companies that reduce their emissions could gain revenues from selling their credits, thus incentivizing emissions reduction of key GHGs. In order to help developing countries participate in the Kyoto Protocol, the UN established an international carbon market under the Clean Development Mechanism (CDM). Countries that already have baseload coal-fired power plants would need to buy emission credits and, if a country like China sold credits, they could afford to build lower emission power generation like gas or wind instead of coal. Since the beginning, in 2003, the CDM has resulted in an estimated reduction in CO_2 emissions of around 1%. Most of the CDM credits have not been used for clean energy but rather have been going to industrial chemical processes, including reducing methane emissions from landfills by burning, chemically destroying HFC-23 produced in refrigerant manufacture and reducing NO_x emissions from fertilizer manufacture (Wara 2007).

The transition management approach for developing policy to achieve sustainability was pioneered in the Netherlands and combines a long-term vision with short-term experimental learning, innovation and risk-taking, and rewards both success and failure (Loorbach and Rotmans 2010). However, the project was abandoned after several years due to political pressure.

2.1.5   INNOVATION NEEDED TO ADDRESS UNSUSTAINABLE ENERGY

Sustainability is quite easy to define: it means that what you are currently doing works, and will keep working, and can be replicated indefinitely. Thus, *unsustainable energy use is a design flaw and will cause system failure at some point.* The primary reason that our current energy systems are unsustainable is the reliance on fossil fuel. The primary source of risk to normal operation is the reliance on oil for nearly all transportation. Some engineering consulting companies are now providing sustainability engineering in supply chain, procurement and logistics for companies, cities and the military (Mathaisel et al. 2013). The current practice of energy management and sustainability engineering uses energy auditing as a main tool – measuring the carbon footprint, energy intensity and costs associated with energy, water and waste disposal. Auditing of current energy use and assessing the exposure of transport activities to fuel price rises is also a good starting point for assessing the risks of reliance on unsustainable energy.

We have seen that regulatory restrictions on harmful activities always occur *after* disasters or failures. In safety or security engineering, these failures can be immediately obvious. Regulations and rules to reduce risks of future failures are enacted after engineering remedies to the problem have been developed. Safety improvements have resulted from innovation, research and development of new products such as eye protection and air bags. But the real innovation in Safety Engineering is the approach of preventing what is preventable. The approach is to identify and analyse accidents and failures in order to change unsafe practices.

Internal combustion vehicles fuelled with petroleum products work very well, unless there is a fuel shortage or unless the emissions accumulate and cause damage. The regulations that limit vehicle emissions caused changes in automobile engineering and reduced smog in cities, but they did not reduce automobile use. Biofuel mandates have been enacted by the US Congress and European countries. Thus far, manufacturing biofuels from food crops has required subsidies and has led to food shortages, while biodiesel from palm kernel oil has resulted in rainforest destruction. There have been a few restrictions on personal automobile use. Shanghai has a limited number of vehicle registrations, and there are congestion restrictions in Singapore and London. However, there have been no restrictions on personal use of fossil fuel in vehicles or in flying. Energy transition to manage fossil fuel risks involves innovations that restrict the amount of fossil fuel produced and consumed.

Let us use a broad view of innovation when looking at energy transition. Transition Engineering innovation includes new businesses, redevelopments or technologies that shift customer behaviour because the use of fossil fuel is restricted. This kind of innovation would be surprising and would be most likely to be done by engineers who already work in systems that produce and use fossil fuels. Politicians do not innovate; rather, they generally react to a crisis and decide between alternatives. Economists do not innovate; they generally collect data on production, consumption and costs and monitor trends. Individuals and households do not innovate; they typically behave according to social norms and use existing infrastructure, energy supplies and technology according to the label. Environmentalists do not innovate; they tend to identify unsustainable practices and work to stop pollution or ecosystem destruction. Scientists do not innovate; they carry out rigorous observation of what is happening and prepare those observations as evidence of causal relationships between industrial activities and environmental and health impacts. It is time for engineers to disrupt the unsustainable pathway.

Transition engineers can use all of the information about the unsustainable activities to feed into new innovative technologies, practices and operations. Energy transition innovations will down-shift fossil fuel use while managing the risks to well-being, quality of life and prosperity. Most important, transition engineers understand that innovation is essential because the main problem is *how to change existing systems*. By definition, innovation can only occur when no existing solution will work. The key element in innovation is a change of perspective; that means seeing the same things in a different way. In this section we have found out that current ideas about sustainability have not worked to change the unsustainable energy activities, and that means that new ideas about how to approach the problem are needed.

2.2 PROBLEMS OF CARRYING CAPACITY AND RESOURCE CONSTRAINTS

In this section, the perspective on growth will shift from being a development goal to being a risk. Elaborate modelling scenarios are not generally needed in order to explore issues of unsustainability. The engineering approach will be based on measurable facts about energy and resource systems. The best approach is to gather historical data from publicly available sources, then use simple models for growth, decline and depletion to explore the characteristics of unsustainability into the future.

2.2.1 THE PROBLEM OF EXPONENTIAL GROWTH

Energy use has grown during the past 150 years in lockstep with growth in population. Energy production growth has also followed similar trends as have water, food and mineral use. Over the same period, forests, wild fish populations, natural wetlands and wilderness have declined. Rationality dictates that continuous growth in a finite system is not possible, and yet, at the same time, we believe growth is necessary and inevitable (Georgescu-Roegen 1976).

We can understand that the economic system will have to respond to the unsustainable resource uses, but we cannot analyse or predict economic behaviour the same way we can analyse or predict physical systems. The simple and clear arithmetic of the exponential function is the most important tool that engineers can use to analyse, understand and communicate the issues associated with past and future growth. Suppose a quantity of goods is consumed in a given year, and the amount of consumption increases each year by the same percentage. The consumption in any future year is given by:

$$N(t) = N_0(1+r)^t \tag{2.1}$$

where:

$N(t)$ = consumption of quantity N in the future year t
r = annual percentage growth rate
N_0 = initial quantity in year $t = 0$

After one year, the consumption will be $N(1) = N_0(1+r)$; after two periods, it will be $N(2) = N_0(1+r)^2$ and so on. The time it would take for the original quantity to double, called the doubling time, is of interest in studies of growth. Setting $N(t) = 2N_0$ in Equation (2.1) and solving for t gives:

$$D = t = \frac{\ln(2)}{\ln(1+r)} \tag{2.2}$$

where:

D = the doubling time

For most practical situations, r is small, say, less than 10%, so we can approximate $\ln(1+r) \cong r$. Noting that $\ln(2)$ is equal to 0.693, Equation (2.2) simplifies to the rule-of-thumb approximation for doubling time:

$$D \cong \frac{0.7}{r} \tag{2.3}$$

Thus, for a growth rate of 7%, the doubling time would be about 10 years.

Now suppose we want to look into the future and determine how long a finite resource will last. First, consider a constant consumption rate of a resource, that is, consumption with no growth ($r = 0$). The remaining lifetime for *constant* consumption rate of the resource would be:

$$t_L = \frac{N_0 - N_T}{N_P} \qquad (2.4)$$

where:

t_L = remaining lifetime of the resource
N_0 = initial quantity of the resource
N_T = amount that has been consumed up until the present
N_P = constant periodic consumption

This is a simple linear relation to evaluate the years remaining at the current rate of consumption. However, when the consumption rate is increasing every year by the same percentage, called the constant growth *rate*, the lifetime of the resource is given by:

$$t_L = \frac{1}{r} \ln\left(\frac{(N_0 - N_T)}{N_P} r + 1\right) \qquad (2.5)$$

Growth is a natural part of all living systems. However, growth is only the first part of the inevitable cycle of growth and decay. When we are considering the carrying capacity of a region or the lifetime of a finite resource, we will use this quantitative analysis to calculate future consumption quantities and lifetimes of resources.

Example 2.2

Between 1960 and 1970, oil production in the United States grew at an annual rate of 2%, from 2,575 to 3,517 billion barrels per year, as shown in Figure E2.1. Using Equation (2.1), we can see that if this trend had continued, by the year 2000, the production would have been 6,371 billion barrels, and the cumulative consumption would have been 222 trillion barrels. In 1970 the upper estimate of proven reserves was 200 trillion barrels. How could the United States run out of oil in just 30 years? It turns out that the oil did not run out; the oil production peaked in 1974 and started declining rather than continuing to grow. In 2000 US production was 2,125 billion barrels. The global oil price spike in 2007 and 2008 set off expansion of production of previously known but uneconomically recoverable reserves of unconventional 'tight oil'. During the tight oil boom, the annual growth rate has been as high as 17%.

2.2.2 THE POPULATION PROBLEM

The world's population has been on an exponential growth path since the eighteenth century (United Nations 2012). Although not thought of as an engineering topic, the challenge of sustainable development is intimately linked to population. Figure 2.1

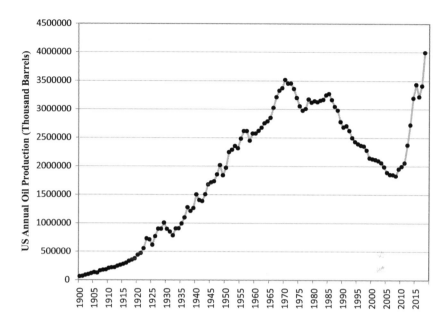

FIGURE E2.1   Historical US crude oil production including liquid condensate, extracted from the Energy Information Administration data (https://www.eia.gov/beta/international/).

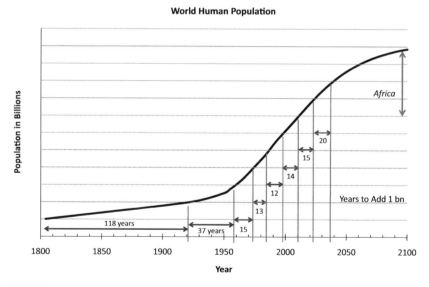

FIGURE 2.1   Global population with number of years to grow by 1 billion, and projection to nearly 11 billion people in 2100, when more than 4 billion will live on the African continent. (From UN 2012.)

shows historical data and growth projections for the world population from 1800 to 2050. Current population is a function of the number of people alive last year, minus the number who died, plus the number who were born. The population of a given area is also a function of the number of people moving in and out of the region. A fertility rate of two children per woman would result in a stable population if the death rate equalled the birth rate.

Throughout most of human history, life expectancy was much lower than today. Human cultures universally developed practices and norms that reflect a high value on fertility, or the number of children per woman. The dramatic increase in world population in the last century was largely due to reduced infant and maternal mortality and improved sanitation, not higher fertility. At the start of the twentieth century, all countries experienced rapid population growth, but with each generation, the number of children per woman has been declining. The rate of population growth has slowed appreciably in many developed countries, but not in Africa, the Middle East and South America. In the United States, the native-born population decreased for the first time in its history during 2009. In western countries, including Italy, Germany and Canada, the percentage of women having children in a given year has dropped to a point where the population is no longer growing, even though the longevity is very high and the infant mortality is very low. Fertility is lower where women have many other socially valued roles besides bearing children.

Demographics, in particular the age distribution, can have an impact on energy use. Japan is an interesting case because the average age of its population has just reached 50 years and is still increasing. Earlier in the chapter, we looked at the exponential growth in energy demand experienced over the past 70 years. A major issue for electricity and fuel suppliers was keeping up with the population growth and the increased demand per person. However, the energy use per person in the United States, Europe and Japan has not grown for more than two decades (IEA 2015). Retired people tend to use less energy than households with young people and children. Thus, the population issue in some countries actually means that the energy system has reached a transition point where demand for electricity and fuels is not trending upward any longer and is in fact declining.

There has rightly been concern about population growth for the past 50 years. However, the projects of the energy transition for rich countries are about their own large-scale reduction in energy and resource use per person. People who have energy-intensive lifestyles are often not comfortable considering reduced energy use, and they get distracted with the idea that people in developing countries are the problem because they want to have the American or European lifestyle. In fact, research shows that people in stable traditional societies primarily wish to preserve their environmental resources, their sovereignty and their way of life (CSI 2012).

Example 2.3

Take some time to explore the World Bank data using Google Public Data (www.google.com/publicdata) and look at fertility, life expectancy, energy use per capita and population growth rate over the past 60 years using bubble graphs. Study different countries that have vastly different energy use per capita in Kg oil equivalent (kgoe) and population growth rates. All African countries other than South Africa have per capita energy use below 1,000 kgoe/yr, and much of that is wood and dung. Nearly all of Asia had energy use well below 750 kgoe/yr in the 1980s, but China and others grew rapidly, to around 2,000 kgoe/yr today. Compare this to Northern Europeans, who consume 4,000–5,000 kgoe/yr, and North Americans, who consume around 7,000 kgoe/yr, the vast majority of which is fossil fuel. We must be clear that the objective for energy transition projects is to achieve dramatic reductions in energy use by companies, organizations and individuals in *developed* countries, where the population growth rate is less than 2%.

2.2.3   UNSUSTAINABLE DEMAND ON FRESHWATER

Less than 1% of the water on the planet is suitable for drinking and agriculture; the rest is saltwater, brackish or frozen. Freshwater refers to rivers and lakes fed by seasonal precipitation. Aquifers are underground freshwater reservoirs in permeable gravels or sand. Unconfined aquifers are replenished by surface precipitation, while confined aquifers, like the massive Ogallala aquifer under the American Great Plains, are actually finite, fossil water, deposited over 1 million years ago. The sustainable water consumption in any given location is constrained by the precipitation rate, storage systems and the requirements to sustain local ecosystems. Confined aquifers have a finite lifetime. In Texas, artesian wells were plentiful in the early twentieth century, but wells must now be pumped and in some places have run dry. A recent study found that 30% of the Southern Great Plains will have exhausted the groundwater resource within the next 30 years (Scanlon et al. 2012).

The World Health Organization (WHO) stipulates that the basic requirement for water is that 20 litre/day per person be accessible within 1 km of the home. In an industrial society such as the United States, personal water consumption is 10 times this level, somewhere between 200 and 300 litre/day for household uses. But if the industrial and energy production usage are added, freshwater usage can exceed 5,000 litre/day on a per capita basis. Water scarcity is defined as annual availability of less than 1,000 m^3/person (World Bank 2003), and 1,200 m^3/person is the minimum for modest consumption (Hoekstra and Chapagain 2008). The population-carrying capacity of a country is calculated by dividing the currently available potable freshwater supply by the consumption per person. Many of the countries that have exceeded their carrying capacity for water use are destabilized, with populations experiencing famine and conflict, and with conditions that result in people leaving their traditional homes and becoming environmental refugees. These water-stressed countries include Egypt, Ethiopia, Jordan, Algeria and Kenya (UN 2015). Some countries with large water deficits use energy to desalinate brackish water or seawater. These countries include South Africa, Israel, Australia, Morocco, Maldives and Middle Eastern countries.

More than 7,000 desalination plants are in operation worldwide; 60% are located in the Middle East. The levelized cost of water produced by various desalination plants is between $0.60/$m^3$ for large plants and $3.00/$m^3$ for small remote plants. The energy requirement varies greatly with the type of process and is much lower for brackish water than for seawater. Thermal processes like multistage flash desalination (MFD) uses 10–15 kWh/m^3 of water, while reverse osmosis (RO) uses 0.5–2.5 kWh/m^3 for brackish water and 4–13 kWh/m^3 for seawater (Loupasis 2002). In Saudi Arabia, the world's largest desalination plant produces 0.5 million m^3/day of potable water. Industrial-scale RO units produce 50–85 litres of water for every 100 litres used, and producing 50–15 litres of brackish water as waste. RO requires energy for a pump that increases the pressure of water on a cross-flow membrane, and only water molecules can pass through the membrane to supply potable water. The water supply must be pretreated by filtration, softening and removal of chlorine biocide with activated carbon. RO membranes are expensive and subject to fouling if the pretreatment and maintenance are not carried out correctly.

Even readily available surface freshwater requires energy for pumping and pressurization (0.14–0.24 kWh/m^3) and treatment (0.36 kWh/m^3) (NRC 2008). Around 3% of the electricity demand in the United States (56 billion kWh/yr) is used for treating and pumping freshwater. California has a large agricultural sector and a large population living far from freshwater sources, resulting in 19% of electricity demand associated with provision of water.

With predicted population growth, an increasing number of countries have substantially exceeded the water-carrying capacity for even minimum survival needs, requiring large amounts of energy for desalination. Without this, social problems erupt and people are forced to leave their homelands (Goldemberg and Johansson 2004). Freshwater is a critical issue for most island nations. Fresh groundwater is available on islands only in a lens that is continuously replenished by rain. Diesel pumps and deep bores brought in to provide reticulated water supplies can easily reduce the pressure in the freshwater lens enough for seawater to infiltrate, as is the case in the Maldives, where all washing and drinking water must now be made from seawater by RO plants, run by electricity from diesel generators. The western states of the United States, from Texas to California, have experienced extreme or exceptional drought conditions for much of the past 15 years. These are conditions that once occurred only rarely, possibly once in a generation (USDA 2017).

2.2.4 FOOD SUPPLY ISSUES AND THE BIOFUEL PROBLEMS

Food production must be adequate to feed the population, and all people must have access to food in order to have sustainable development. The Earth has an estimated 3.32 billion hectares (8.2 billion acres) of potentially arable land, divided according to the following levels of productivity (Barney et al. 1993):

Highly productive	445.5 million ha
Somewhat productive	891 million ha
Slightly productive	1,985 million ha
Currently productive	1,458 million ha

In the United States overall, about 0.6 ha of highly productive land is used to produce the food for one person per year, but estimates of food wastage in the United States are as high as 40% (USDA 2015). A more modest per capita global footprint for diets with less meat and more grain is estimated to be about 0.4 ha per capita. The global carrying capacity for food would be 2.2 billion people using 0.6 ha of only the highly productive and somewhat productive land. Using all of the potentially arable land and a modest diet, the world could support 8.3 billion people. The global population is near the carrying capacity. Currently, one in seven people living in Asia and Africa are chronically malnourished in either total calories or protein (Associated Press 2009). Mechanized tilling degrades soil depth and quality, reducing land productivity. Agricultural land loss to urbanization is in the order of 0.6 million ha per year in the United States.

Clearly it is possible to have food production systems that do not rely on fossil fuel, and much of Asia, Africa and South America primarily use human and animal energy. The food supply system in the United States and other industrial countries requires fossil fuel. The current American food supply system requires approximately 2,000 litres of oil equivalent per year per person, and accounts for 19% of the total national energy use, with 14% for farming, processing, packaging and storage, and 5% for transportation (Pimentel et al. 2008). Corn, soybeans, wheat and sorghum are the basis of the American food production system. In 2000 the United States produced about 50% of the world corn and soybean supply, 25% of the wheat supply and 80% of the sorghum supply. In the United States, about 12% of the domestic corn market is for food; the majority of corn is used for animal feed and ethanol.

Biofuel has become an international food issue. Ethanol production grew rapidly from 2005 in response to the US Renewable Fuel Standard of 2007, which established government mandates for refineries to blend increasing amounts of biofuel into fuel products. Ethanol production was 796 million litres in 1980 and 64 billion litres in 2011. As a consequence of the bioethanol boom from 2000 to 2011, the percentage of corn crop used for food in the United States dropped from 13% to 11%, exports dropped from 20% to 13%, and feed for US cattle and chickens plummeted from 60% to 36% of the annual crop. In 2011 ethanol processing consumed 40% of the total US corn crop, and the monthly farm price for corn hit an all-time record, three times higher than in 2006. The US government mandated continued growth in biodiesel supply, to over 160 billion litres by 2022, but in 2014 the ethanol production fell short of the mandate of 65 billion litres, and growth in corn ethanol is not expected to be possible into the future. The government policy was based on anticipated commercial cellulose or algae ethanol beginning production by 2012, which has not happened.

Biodiesel production in the United States expanded from 2.3 million litres in 1999 to 3.6 billion litres in 2011. Despite this rapid growth, biofuel in total accounted for 6% of US transport fuel supply on a volume-equivalent basis in 2011 (Schnepf 2012). The European Union (EU) uses much more biodiesel than ethanol as diesel passenger cars are common. Germany and France together produce 50% of Europe's 8.6 million tons of biodiesel. The current production uses around 3 million ha of land in Europe. But the food issue associated with Europe's biodiesel is actually in Africa, where European companies buy up land to produce fuel for Europe instead of food for Africa, accounting for 18.8 million ha of land conversion (Amigun 2011). Mandated use of 10% biodiesel in Europe has also contributed to deforestation and the associated increased CO_2 emissions in Malaysia and Indonesia, where palm kernel oil is produced.

Example 2.4

Competition between agriculture and water for food versus biofuels has become an issue that has effectively derailed support for what are called first-generation biofuels – bioethanol and biodiesel made from edible food crops. The issues are sustainable land use, water allocation and social responsibility. Use the given data to explore the implications of the US Energy Independence and Security Act of 2007, which requires 136 billion litres of biofuel per year to be blended into the nation's gasoline supply (http://www.afdc.energy.gov/fuels/laws/BIOD/US).

Given Data: It takes 15,150 litres of water to grow 1 bushel (25.4 kg) of corn, and an average yield of corn per acre is about, 5,000 kg, grown from about 6.8 kg of seed corn. It takes 3.2 kg of corn to produce one litre of ethanol, and the ethanol production process also consumes 3 litres of water per litre of ethanol. The average household consumes about 760 litres of water per day for domestic uses. A Ford Explorer used by a typical American family has an 85-litre capacity fuel tank, and annual fuel use per household has recently fallen to the level of the mid-1980s at about 3,785 litre/yr (http://www.umtri.umich.edu/).

Analysis: The biofuel mandate corn inputs would be:

$$(136 \times 10^9 \text{ lit(e)})(3.2 \text{ kg corn/lit(e)}) = 435 \times 10^9 \text{ kg corn}$$

The water requirement for growing corn and producing ethanol is:

$$[15,150 \text{ lit(w)}/25.4 \text{ kg}](3.2 \text{ kg/lit(e)}) + 3 \text{ lit(w)/lit(e)} = 1,911 \text{ lit(w)/lit(e)}$$

Substitution of ethanol for gasoline for a typical household would require:

$$(3,785 \text{ lit(e)/yr})(1911 \text{ lit(w)/lit(e)})/365 \text{ day/yr} = 19,824 \text{ lit(w)/day}$$

This water used for fuel is 26 times more water than for all other household uses.

2.2.5   THE PROBLEM OF LIFESTYLE EXPECTATIONS

Unlike food and water, the basic level of energy to sustain a given human population at an acceptable standard of living is more difficult to quantify. Lifestyle can cause huge variations in energy use even between people in the same region. The United Nations proposed a measure of standard of living, termed the Human Development Index (HDI), based on various criteria (Goldemberg and Johansson 2004). The United States and Iceland have the highest energy consumption per capita, but their human satisfaction level is not much different from other developed countries such as France or Germany, which consume about half the energy per capita. The United Kingdom and France have twice the energy economic efficiency as the United States but the same gross national product (GNP) (World Bank 2003). Of course, more energy is used by affluent people, and affluent societies have high lifestyle expectations. But the relationship between energy use and standard of living, health, happiness, education and other measures of a good life is not linear; rather, it is a step up and a plateau.

Imagine that you are preparing for a research trip into an area of the Amazon in Guyana, as shown in photos of the author's research in the village of Kabakaburi (Blair 2019) in Figure 2.2. When you are preparing your pack for the adventure, what energy resources will you include? Guyana is in the equatorial region of South America, so would you take an air conditioner? The answer is obviously, no – you would take a wide-brimmed hat, water purification equipment and light clothing.

What energy services are essential enough that you would carry them?

Your cooking energy needs depend greatly on your menu, so you would make low-energy choices. You would not expect to have a hot shower or a cold beer if you had to carry all of your own energy. The essential energy end-uses would be a flash-light and possibly a communication or navigation device, both of which have very low power requirements that can be supplied by portable batteries. On an extended excursion from civilization you might take along a hand-winding generator to charge your batteries. The gear and technology you would take on your adventure would be manufactured using energy for extraction, fabrication, transport and retail.

The point of this thought exercise is to illustrate the relationship between energy use by individuals and the well-being provided by access to essential energy services. Having no access to any form of energy would pose risks to survival for individuals. If your normal uses of energy are interrupted, you either become stressed or you change to a different system you have prepared. This ability to adjust your

FIGURE 2.2 The village of Kabakaburi in the hinterland of Guyana has a high human development index and resilience, but virtually no gross domestic product (GDP), energy consumption or growth.

expectations, technology and the energy resources you use and still carry out your essential activities is called *resilience*.

Even modest access to energy produces a large improvement in indicators of living standard such as the percentage of children that are underweight or literacy rates. One light in the kitchen area improves sanitary handling of food and reduces disease. One light allows people to read after dark. Enough electricity to charge a cell phone or run a sewing machine or hand tools can support a cottage industry. As a rule, societies with annual energy use of at least 200 GJ/per capita (32.7 boe, or 55.6 MWh) have a high HDI (Lambert 2014).

2.3 WATER AND LAND REQUIREMENTS FOR ENERGY PRODUCTION

2.3.1 WATER USE FOR ELECTRICITY GENERATION

The biggest user of freshwater from lakes and rivers is for cooling in thermoelectric power plants: 556 million m^3 withdrawn and 14 million m^3 consumed per day in 2005 (Carney et al. 2008). Thermoelectric power plants need cooling water to operate the steam Rankine cycle. Much of the water used by a power plant is returned to the river or lake, albeit warmed by 4°C–10°C as allowed by permits to protect the aquatic ecosystem. However, 3% of all US water consumption is evaporated in the cooling towers in power plants and lost as water vapour in the atmosphere. Figure 2.3 shows a breakdown of the freshwater withdrawals and consumption for different end-uses (ASME 2011). Seawater is used to cool power plants, but special considerations need to be made for corrosion and barnacle growth in pipes.

The amount of water consumed or withdrawn depends on the kind of cooling system used for the condenser. In coastal locations or where a large river or lake is available and where the water temperature is cool year-round, the withdrawn water

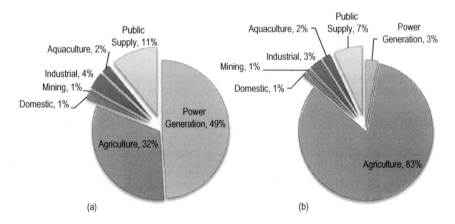

(a)   (b)

FIGURE 2.3 US freshwater withdrawal in 2005 was 410 billion gallons per day: (a) freshwater withdrawals, (b) freshwater consumption distribution.

is returned to the freshwater body. In hot weather, the freshwater is sprayed over evaporation pads, air is blown through the cooling tower, and the evaporation cools the water stream. Thus some of the freshwater is consumed and exhausted to the atmosphere.

Table 2.1 gives the amount of water withdrawn and consumed by various types of thermoelectric power plants. The US average water consumption for thermoelectric power generation was about 105 litres per kWh in 2005 for all types of cooling (Kenny et al. 2005). Water use in electricity generation is an issue of long-term sustainability for all types of thermal vapour cycle power plants, including geothermal and solar (US DOE 2006). There are currently no coal-fired power plants with technology to separate or capture the carbon dioxide from the emissions. If the pulverized coal power plants operating in the United States were to use a monoethanolamine (MEA) absorption process, then the parasitic power would be about 30% of current capacity, and the water withdrawal for coal would increase from the current 27.3 billion litres to 46.8 billion litres (Carney et al. 2008).

Geothermal power plants use hydrothermal resources that are seldom above 300°C. Depending on the pressure of the resource, some of the geothermal brine is flashed and expanded through a steam turbine. The remaining brine and condensate from the steam turbine provide the boiler heat for a binary organic Rankine cycle (ORC), so-called because the working fluid is a hydrocarbon, normally pentane or butane. Organic fluids, ammonia and refrigerants boil at much lower temperatures than water and also condense at higher temperatures. ORCs are also used for solar thermal and waste heat power plants.

Hydropower generation uses the flow of freshwater to produce electricity without consuming or withdrawing water, although the storage lakes do increase the evaporation losses from the watershed. Hydroelectricity supplies about 7%–10% of the electricity in the United States, at the lowest average generation cost of all sources at $0.02 per kWh (US Census Bureau 2007). Dams have been built on rivers where a suitable site provides for a high, narrow dam on solid bedrock, a storage lake and a sufficient height for driving the flow through the turbines. As a rule of thumb, it takes about 14,000 litres of water, with lake storage area of 75 hectares (185 acres) to produce 1 MWh of electricity.

TABLE 2.1
Water Consumption for Different Types of Power Plants

Plant Type	Water Consumption (L/MWh)
Solar thermal	3,182–4,546
Nuclear	3,182–3,637
Subcritical pulverized coal	2,300–2,400
Supercritical pulverized coal	2,000–2,100
Integrated gasification combined cycle (IGCC)	1,400–1,500
Natural gas combined cycle (NGCC)	800–900
Geothermal power plant (air cooled condenser, brine reinjection)	0–10
Waste heat power plant (air-cooled condenser)	0.0

2.3.2 WATER USE FOR FUEL PRODUCTION

Production of fuels also places pressure on water resources, as shown in Table 2.2. Corn ethanol consumes anywhere from 13 to 333 liters of water per liter of Fuel for agriculture and processing depending on irrigation requirements (Wu et al. 2009). There are no commercial processes for producing cellulosic ethanol from wood, but the experimental biochemical processing currently requires 10 lit(w)/lit(f). Biodiesel processing does not require water, but soybean crops are more heavily irrigated, resulting in a water mileage figure of 34 lit/km (12 gal/mi) for driving on biodiesel.

Hydraulic fracturing can require more than 1.5 million litres of water per well for coal beds and 9–23 million litres per well for shale formations. Much of the fracking fluids return to the surface, but they are contaminated and must be disposed of by injection into an approved disposal facility. Some companies are reusing the contaminated producer water for subsequent well fracturing before disposal. Some of the water used to produce gasoline from tar sands is heated and used as steam, with overall water use of 2.6–6.2 lit(w)/lit(f) oil. Water pollution is a major sustainability issue with tar sands extraction operations (Wu et al. 2009).

2.3.3 LAND AREA FOR ELECTRIC POWER PLANTS

Land has ecosystem value and a potential for producing food, fibres and wood. Fossil fuel sources have a high energy density, while all of the renewable sources are diffuse and therefore require large collection areas. Coal power consumes land for opencast mines and occupies land for coal storage, the power plant and ash disposal. The land intensity for coal varies from 150 to 300 W/m^2 depending on coal heating value, power generation efficiency, and the depth of the coal seam.

Conventional oil and gas fields disturb the land where drilling and pumping platforms, pipelines and refineries are located, often creating pollution from spills and

---

TABLE 2.2
Water Consumption for Different Types of Fuel Extraction and Production

Fuel and Resource Used	Extraction or Growing (litres water/litre fuel)	Processing (litres water/litre fuel)
Corn ethanol	10–133	3.0
Cellulosic wood ethanol	–	10.0
Biodiesel (soybeans)	275	0.0
US conventional petroleum	2–6	1.5
Saudi conventional petroleum[a]	1.4–4.6	1.5
Tar sand oil	2.6–6.2	1.5

Source: Wu, M., et al., *Consumptive Water Use in the Production of Ethanol and Petroleum Gasoline*, Argonne National Laboratory, Argonne, IL, 2009.

[a] Saudi oil production requires *desalinated* water.

TABLE 2.3

Land or Water Area for Different Renewable Electric Power Generation Systems

Power per Unit Land or Water Area	W/m²	Availability	U(%)
Wind	0.5–2	Intermittent	24–32
Offshore wind	3	Intermittent	24–32
Tidal barge	3	Periodic	20–40
Solar PV panels	5–20	Intermittent, periodic	10–20
Biomass plantation	0.5	Seasonal	1–2 harvest/year
Hydroelectric reservoirs	1–10	On demand within limits	80–90
Hydroelectric run-of-river	50–800	Baseload	90–97
Geothermal binary and steam	20–400	Baseload, on demand	85–90
Concentrating solar power (desert)	10–15	Periodic	20–30

leaks, but the energy density, even considering 45%–60% conversion efficiency for combined cycle power generation, means that the land intensity for natural gas is in the thousands of W/m^2.

Renewable resources require more land than fossil fuels, but they also can be integrated into the land use, such as solar photovoltaic (PV) on buildings and wind turbines on farms. Table 2.3 shows the power generation per unit land or water body area for various renewable energy technologies.

Hydropower usually involves flooding of valuable river valley land by damming the river to create a storage reservoir. For example, the massive Three Gorges project in China displaced over 1 million people, created a reservoir that consumed 632 km^2 of land, and is designed to produce 18,000 MW. The ability of the supply to match demand is called the utilization factor (U), which is the actual electricity produced by the plant compared to the rated capacity times the hours in a year. For example, a 2 MW wind turbine that produced 5,080 MWh during a given year has a utilization factor of U = 29% because it would have produced 17,520 MWh if it ran continuously at full capacity.

2.4 THE PROBLEMS OF MINERAL RESOURCE DEPLETION AND ISSUES WITH RECYCLING

If you bring up the issues of resource constraints, you will often get this response: 'The Stone Age didn't end due to a lack of stones.' Some archaeologists suggest that human hunters developed a method of hunting by running whole herds off cliffs and into traps, and thus the Stone Age ended because a lack of wild game meant the end of a way of life. Energy engineers find that most people do not like to discuss depletion issues. The cultural narrative is that technological progress will always result in new resources being brought into the market. The transition to sustainable energy systems will most likely be similar to past transitions in that it involves both changes to manage issues and to take advantage of opportunities. Understanding the nature and timescale of the pressures for change is necessary for timely and economical change management and the development of new technologies, redevelopment of old buildings and investment in new transportation systems. There is no benefit to be gained by ignoring a problem until it becomes a crisis.

2.4.1 MODELLING OF DEPLETION

Consumption of a finite resource usually has an exponential growth phase, a production peak and subsequent decline. A well-known model for the production curve for oil, called Hubbert's curve, was developed by M. King Hubbert, and it can be used to model the annual production of a finite resource over its lifetime (Deffeyes 2001):

$$N(t) = \frac{\omega N_0}{\left\{ \exp\left(-\frac{\omega}{2}(t_p - t) \right) + \exp\left(\frac{\omega}{2}(t_p - t) \right) \right\}^2} \tag{2.6}$$

where:

t_p = year of peak production

ω = a shape parameter that adjusts the width for the total lifetime of the production curve

N_0 = the total resource, or estimated ultimate recovery for an oil reservoir

Equation (2.6) is also known as the production curve or depletion curve. Equation (2.6) can be used to gain quantitative understanding of the lifetime of the resource under different production scenarios.

Example 2.5

Consider that a forest tract with 100 million tons of old growth greenheart trees is to be opened up to logging in Guyana. The extraction rate is expected to follow a Hubbert curve as the initial production is slower, then more operators move in with bigger equipment. As the land is cleared, the production slows as bigger operators leave the area. Use Equation (2.6) to look at the difference between a rapid boom and bust that has peak production of 4 million tons at year 30 and a longer cycle with maximum production of 2 million tons at year 50.

Analysis: Figure E2.2 gives plots of the two production boom-and-bust cycles using $\omega = 10$ for the short cycle and $\omega = 20$ for the long cycle. We can see that both scenarios would result in depletion of the trees. Greenheart grows back from a stump if enough of the base is left when the tree is felled. Regeneration to harvestable size requires more than 40 years. Thus, the longer timeframe curve should be the strategy for a sustainable logging industry so that the initially cut trees will be ready for harvesting before the production reaches 2 million tons per year.

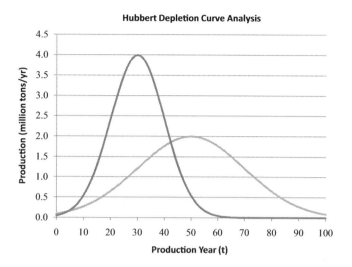

FIGURE E2.2   Example of two different depletion curves, both have total resource extraction of 100 million tons.

2.4.2 MINING

People have been mining for salt, stones, clay and pigments for more than 10,000 years. The Hallstatt salt mines in Austria show archaeological evidence of underground extraction from 5,000 BC. Mining is always dangerous work, but it produces valuable materials that can be traded. The salt extracted from Hallstatt generated great wealth as salt was extremely valuable and transportable. The salt deposit is still being extracted today. As far as we know there are no environmental disasters from ancient mining operations. However, since the industrial revolution, the ability to blast, move and crush vast quantities of ore has left toxic tailings piles and acid mine draining issues all over the world. The rocks, gravels and soils that are exposed to the biosphere are weathered and depleted of water-soluble minerals.

Underground mines will always fill with water, and the oxidation of metal sulphides such as iron pyrite reduce the pH, which then also dissolves toxic metals like copper, zinc and nickel. There are bacteria that accelerate the oxidation process, and the exposure of rocks through crushing and placing in tailings piles greatly accelerates what is normally a slow weathering process in natural formations. Acid mine draining has a characteristic yellow-orange colour and is detrimental to aquatic life. Hard rock mining operations from the 1800s are still causing environmental disasters such as the Gold King Mine (abandoned 1923) blowout in Colorado, the United States in 2015. The accumulated acidic and toxic water in the mine blew out into the Los Animas River, creating a bright orange plume that progressed through New Mexico and Arizona. The water was contaminated with iron, lead, cadmium and arsenic at concentrations hundreds of times above safety limits. Coal mines can have acid drainage but can also produce alkaline drainage if the coal was co-deposited with limestone.

Another increasingly common environmental problem with mines is the failure of dams built to store mine tailings. Ores extracted from mines are 'milled', usually crushed and pulverized to expose more surface area for chemical extraction. Mill tailings are the crushed ore rocks that have undergone hydrometallurgical chemical leaching to extract the desired mineral. For example, low-grade gold ore is crushed and mixed with water and sodium cyanide or calcium cyanide, which binds to the gold. After the desired minerals have been extracted, the tailings are toxic waste that can easily become fluid in water or liquified in an earthquake and will continue to leach acid and heavy metals into water. Tailings are put in a pile or left to settle out of processing water in a pond behind a low embankment dam.

There had been 147 major disasters involving tailings dam failures as of 2008. The 2015 Samarco iron mine disaster in Brazil is an example of extensive and near complete destruction caused by the huge volume of toxic material that can become liquefied and thus produce plumes of toxic sludge flowing through valleys and down rivers. There is currently no way to stop such a disaster once the tailings dam fails. There is also no long-term solution to the problem of aging dams and unstable piles at mines where the extraction company has long closed up operations and gone into bankruptcy (a favoured way to deal with environmental liabilities) (Rico et al. 2008).

Mining and milling of materials has long-term implications for energy consumption because all future mineral extraction and production will consume more energy per unit production than in the past. The amount of energy and water required to extract virgin minerals is a strong function of the depth of the deposit, the ore grade (the percentage of the mineral in the ore) and the type of native ore. The ore grades for all materials have been trending steadily downward since records began (Prior et al. 2012). The technology and modelling in geological investigations has become increasingly sophisticated. The known mineral deposits of nearly all industrial minerals have not substantially increased for many years, but the amount of ore considered as *reserves* has been increasing. This strange phenomenon is due to the fact that reserves are determined by the price of the mineral. As high-grade ore deposits are consumed and if demand increases and the price increases, then lower-grade ores become economically recoverable reserves (Mudd et al. 2013).

The extraction and processing of the resource and bringing it to the market are a function of economic factors and the regulatory environment. Consider the beginning of the extraction of copper in the 1920s, when it started to be used as an electrical conductor and for motor windings. Copper production initially grew, leading to growth in demand. The production cost usually declined as production expanded and mining and extraction technology improved. The tailings from milling and leaching were simply piled nearby. There were no facilities or systems for recovery and recycling, and recycling would have been more expensive than primary production. As demand grew, more mines opened. Mining technology got even bigger and better, reducing production costs. The increased supply kept the copper price low. However, as the peak of the production curve is reached, the demand is high and growing, but supply cannot grow, and price climbs to the point where recycling becomes profitable. As the primary mined copper supply declined, recycled copper began to represent a new and significant production stream in the market. Also, as new technologies are developed that need the rare earth elements discarded during the copper production phase, the tailings may then be reprocessed using increasingly aggressive chemical separation processes.

Large-scale changes in the demand for a given mineral can also reduce the production rate. For example, primary production of lead mining has decreased due to regulations banning lead in paint, lead gasoline additives and electronics solder, together with regulations requiring recycling of batteries. Long before the last ton of ore is extracted from the mine, the demand for the diminishing, and thus increasingly costly, resource will have also greatly diminished. Production curves do not follow smooth curves, but all extracted resources, including renewable resources like forests and fish, have followed the general pattern. We can't know for certain what the future economic, technology and consumer demand situation will be, but we can use the depletion curve to understand that future growth in demand is not possible past the production peak.

2.4.3 CRITICAL MATERIALS FOR RENEWABLE ENERGY AND EFFICIENCY

Wind and solar energy are not finite or subject to depletion effects, unless we consider land limitations. However, some of the critical materials, known as rare earth elements (REEs) are presenting challenges to expansion (Massari and Ruberti 2013). Table 2.4 gives some of the key rare earth elements, their uses and the price over a 3-year period.

The eight elements in Table 2.4 account for 90% of all the REEs used. These elements have about the same abundance in the Earth's crust as lead, and are about

TABLE 2.4

Use of Rare Earth Elements in Alternative Energy Conversion, Air Pollution, Fuel Processing and Energy-Efficient Technologies with the Price Rise over 3 Years Resulting from Supply Insecurity

Rare Earth Element	Uses in Energy-Related Technology	2009 $/kg	2010 $/kg	2011 $/kg
Lanthanum (La)	Lighter and longer-lasting batteries, catalysts for petroleum cracking, phosphors	6.25	60.00	151.50
Cerium (Ce)	Efficient display screens, high activity catalysts, lighter and stronger aluminium alloy, phosphors for fluorescent and halogen lamps	4.50	61.00	158.00
Neodymium (Nd)	High power, permanent NdFeB magnets, catalysts	14.00	87.00	318.00
Praseodymium (Pr)	Magnets for hybrid/electric vehicles, wind turbines, fibre optics, catalysts and phosphors	14.00	86.50	248.50
Samarium (Sm)	Magnets for hybrid/electric vehicles, wind turbines	3.40	14.40	106.00
Dysprosium (Dy)	Magnets for hybrid/electric vehicles, wind turbines	100.00	295.00	2,510.00
Europium (Eu)	Magnets for hybrid/electric vehicles, wind turbines	450.00	630.00	5,870.00
Terbium (Tb)	Magnets for hybrid/electric vehicles, wind turbines, phosphors for fluorescent and halogen lamps	350.00	605.00	4,410.00

1,000 times more abundant than gold. Currently, more than 80% of these elements are produced from one mine in China, which set export quotas in 2010 to preserve supplies and curtail environmental pollution. This resulted in shortages and price spikes in 2011. Rare earth mineral prices have since collapsed by on average 70% in 2012 to about three times the 2009 price, due to a range of factors, including release of privately held reserves. In 2014, the World Trade Organization (WTO) ruled against the Chinese export quotas and tariffs, stating that while the efforts to manage limited resources and environmental impacts were understood, China's actions constituted unfair trade practices.

Lanthanum (La) and cerium (Ce) are the most common of the rare earth elements. Ce is used in water purification to remove phosphorous pollution from agriculture. La is used in rechargeable batteries to adsorb hydrogen. La is used in hybrid car batteries, Ce is used in catalytic converters, and both are used in gasoline reforming. Hybrid/electric vehicles and wind turbines require high power magnets containing Nd and Dy to make motors and generators smaller and lighter.

Single-crystal and solar PV panels currently rely on ultrahigh-purity silicon, and small amounts of dopants including gallium, arsenic, phosphorous, and boron. Efficient light-emitting diode (LED) lighting fixtures also require these dopants, and compact florescent lamps (CFLs) require mercury and phosphors. Videoconferencing can replace a large amount of fossil fuel use in travel, but manufacturing computers requires nearly every element in the periodic table, including the REEs Eu, Y and Tb. Unfortunately, some of the key materials vital to renewable energy and energy efficiency are not renewable, abundant or low impact.

The mineral resources for REEs are not rare in the sense of total quantity, but they are found in a limited number of deposits in high enough concentration that can be economically mined and processed. Extracting REEs from mineral ore presents serious technical and environmental challenges. The ore is first milled (crushed to fine particles), then it undergoes a series of processes involving numerous chemicals, acids and reagents, producing highly toxic and radioactive wastes. Very serious pollution issues have occurred in the places where REEs have been mined and milled, for example, at Mountain Pass, California, and in China.

Demand for REEs for magnets in motors was expected to increase by 10%–15% from 2010 to meet the growing demand for hybrid vehicles and wind turbines. A small amount of REEs in iron increases the magnetic power, thus greatly reducing the size and weight of the electric motor. Most of these elements currently have no known substitute or potential for recycling. It is important to note that even the supply of a material that is used in only trace amounts can represent an issue for energy technologies.

2.4.4 RECYCLING OF MINERALS

Recycling is one of the first sustainability initiatives in most organizations. *Recycling* is the term used for a whole chain that starts with diverting discarded objects from going to the waste disposal, then separating the different types of objects (e.g., paper, glass and metals), then recovering the useful component materials by further separation and chemical and heat treatment. In fact, recycling is a manufacturing industry that produces feedstock pulp, ground glass and plastic beads and fibres.

For many minerals, like copper, aluminium and lead, the end-of-life recovery from products is now more economical, and less energy intensive, than extraction from any new mine deposit. Figure 2.4 was prepared by a mineral geology professor to model the future outlook for copper – an absolutely critical material for energy systems (Vidal 2015). The depletion effect for primary copper is clear, but copper can be recovered from waste in wires, motors, tubing, heat exchangers and so on and reused. The percentage of copper currently recycled is low, but that is primarily because it is still in use. As equipment, buildings and products start to be retired, secondary copper could become the main source for new products after 2050. The model in Figure 2.4 assumes that copper in service becomes available for recycling after 30 years. The model also assumes a steadily increasing recycling rate of retiring copper from the present 35% to 75%. A finite, recyclable resource could potentially provide for a certain level of use in new products, but it cannot provide for continued growth. Companies or countries that transition their operations to design for recyclability and set up collection and recovery systems will be positioning themselves to have access to materials into the future. Note, however, that because 100% of metals cannot be recovered and recycled, there is eventually a decline in the supply.

People are living in closer proximity to mining operations and are becoming more sensitive to the long-term environmental effects, and less tolerant of the health risks. Regulatory approval of new mines is becoming more difficult, mine operators are required to spend more on mitigation, and reprocessing of previous milling waste is increasing (Prior et al. 2012). Mineral resource depletion for all materials

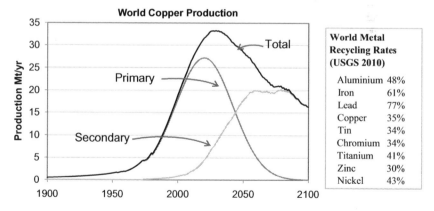

FIGURE 2.4 Future growth in copper production is not possible, continuing availability of copper will increasingly depend on secondary (recycled) copper, and the energy consumption for primary copper from mining will become prohibitive. The graph illustrates future supply plateaus expected for all metal materials. (From USGS, Mineral commodity summaries, Technical Report, US Geological Survey, http://minerals.usgs.gov/minerals/pubs/mcs, accessed February 2015, 2010.)

should now be considered one of the mega-issues driving transition to recovery of materials and design for reduced use of materials. As with oil, depletion of minerals is not a matter of depletion, but it is a matter of restricted supply, increase in cost and upswing in recycling operations.

Reducing use, reusing and recycling are fundamentals of design for sustainability. The opportunities for recovering and recycling materials are sensitive to the product design and establishing partnerships with regulators and retailers. The cost of separating the different materials in a product can be greatly reduced if the end of life is a consideration in the design and manufacture. For example, glass bottles can easily be returned and reused for the same product if the labels are either removable paper or etched on. Recycling glass by crushing it, dissolving off the labels and melting for blowing is much more energy intensive. Thin-film materials and doped semiconductors used to fabricate all of our electronics and solar panels are virtually non-recyclable.

Worldwide the most important and valuable metals are being recycled to some degree (USGS 2010). There are currently industries that collect, separate and reprocess the main recycled products in developed countries. The EU has a mandated target of 50% recycling of household and commercial waste by 2020, and overall rates are currently at 35%. Austria (63%) and Germany (62%) have strong laws requiring manufacturers to take back products at the end of life for recycling. This law plus waste collection requirements have led to transitions in product design, in the materials used, assembly and manufacturing. None of the REEs used in electronics and solar panels are currently recycled. Recovery and recycling are very difficult for elements used as dopants, as micron-scale connectors (e.g., silver), and as nano-materials (e.g., ruthenium and platinum).

2.4.5 ENERGY AND MATERIALS FOR BATTERIES

Batteries store energy for mobile applications, and public interest has been high in utility-scale battery energy storage (BES) facilities to manage intermittent production from solar and wind and to meet peak demand. Batteries will be covered in more detail in later chapters, but the main unsustainability problems of batteries are the fact that they consume finite materials, often use hazardous materials, require energy to produce and incur an energy penalty when used. Storing electricity in a battery requires two conversions from electricity to chemical potential, then back to electricity. Because most electricity production and use are AC, there may also be inverters used to convert AC to DC, which must be used for all batteries. Energy conversion and rectification/inversion consume energy. Using a battery incurs an energy penalty of 10%–40%. Everyone will be familiar with the one-time use of non-rechargeable batteries, and the finite number of useful charge cycles for rechargeable batteries. Lanthanum is a rare earth material now widely used in lithium batteries. Table 2.5 gives some energy values from cradle-to-gate assessments carried out since 1979 for battery materials (Sullivan and Gaines 2010). Lithium-ion batteries are not currently recycled. The energy and cost of materials is also important because the energy used to extract and manufacture things like batteries and wind turbines would be produced from fossil fuels.

TABLE 2.5

Production Energy from Lifecycle Analysis for Cradle-to-Gate Production Energy Needed for Raw Materials Extraction from Ore or Recycled Products and Delivered in Bulk for Manufacturing Batteries

Lead-Acid Battery Materials (Specific Energy 20–50 Wh/kg)		Lithium-Ion Battery Materials (Specific Energy 75–120 Wh/kg)	
Material	Production energy (MJ/kg)	Material	Production energy (MJ/kg)
Pb – virgin	22.3–31.2	Co precipitation	144
Pb – recycled	4.2–12.7	Li_2CO_3 from brine	44.7
Polypropylene – virgin	73.4–75.5	$LiOH–H_2O$ from ore	163
Polypropylene – recycled	15.1	LiCl from ore	220
Glass	20	Graphite	187–202

Source: Sullivan, J. L. and Gaines, L., A review of battery life-cycle analysis: State of knowledge and critical needs, Argonne National Laboratory, ANL/ESD/10-7, Oak Ridge, TN, 2010.

Example 2.6

Consider the proposition of utility-scale battery energy storage for wind generation. The Western Interconnection grid in the United States has total installed capacity of 265,000 MW and supplied 88,360 GWh to consumers in 2015. Peak winter demand is 120,000 MW over 2 hours, and is supplied by gas with peak price $0.268/kWh (WIPG USA 2016). Sodium-sulphur batteries have been used for grid-scale storage in Japan. The round-trip storage efficiency is 75% and the maximum discharge depth is 33% for the longest battery cycle lifetime of 7,200 cycles. They must operate above 300°C and have a molten sulphur cathode, solid alumina electrolyte and molten sodium anode. The battery is 22% aluminium, 13% steel, 10% alumina, 8% Na and 12.5% S, and 15% sand for insulation around the battery. The specific energy of the battery is 116 Wh/kg and cost is $320/kWh (Sullivan and Gaines 2010). If 2% of the peak demand were to be supplied by storing off-peak power for discharge on peak, explore the storage needed. For reference, the US sulphur production in 2013 was 9,210,000 kg, mostly recovered from stack scrubbers.

2% of peak demand = 120,000 MW × 2 hr × 0.02 = 4,800 MWh
Storage capacity = 4,800 MWh ÷ 0.33 = 14,545 MWh ÷ 0.75 = 19,394 MWh
Na/S battery weight = 19,394 × 10^6 Wh/116 Wh/kg = 167.2 × 10^6 kg
S weight = 167.2 × 10^6 kg × 0.125 = 20.9 × 10^6 kg (~2 times 2013 US production)
Na/S battery cost = 19,394,000 kWh × 320 $/kWh = $4.654 billion
Cost of storing electricity = $4,654M ÷ (4,800,000 kWh × 7,200 cycles) = $0.135/kWh
Cost of electricity stored = $0.268 ÷ 0.75 = $0.357/kWh
Embedded energy = 125,391,850 kg × 235 MJ/kg ÷ 3.6 × 10^6 MJ/GWh = 8,185 GWh

We have all heard things like 'We need batteries to store renewable energy to use for peak demand.' The Western Interconnection grid in the United States reports the average utilization factor for wind in the region is 0.21. In the above calculation of Na/S storage capacity for 2% of the winter peak demand, the required invest-ment in wind-generated electricity, assuming sufficient generation is achieved each day, the wind capacity would be = 4,800 MWh / 0.75 / 24 hr / 0.21 = 1,270 MW. This would make it the second largest wind farm in the United States. In addition to the cost of the batteries above, wind farm development capital cost is between $1,700 and $2,450/kW. Thus, adding the wind capacity and battery storage to supply 2% of the area peak demand would cost at least $6.8 billion.

Now, let's flip the perspective to the demand side and see what could be done with $6.8 billion to reduce peak demand. The technology-driven scenarios are easy to conceive, and easy to analyze to understand the order of magnitude of such a system compared to the current system. In this example of wind-battery storage, we might forget that the actual problem was that peak demand requires investments in grid infrastructure and peak generation, which is from gas-fired power plants. The addi-tion of batteries and wind turbines would require more grid investment. The winter peak demand would be occurring during the dark and cold winter evenings. With a population of roughly 120 million people, this peak demand shaving would amount to = 2,400 MW/120 × 10^6 = 20 Watts per person. This is less than the power required to charge a cell phone. Doesn't it seem strange that demand side reduction measures don't receive the same attention as energy technology development ideas?

Demand side management (DSM) refers to investing in changes in the end-use appliances and operations to make the electricity supply more affordable or more reliable. In this example, the cost per unit of electricity that is stored and then delivered is $0.135 + $0.357 = $0.492/kWh. A Transition Engineering project that reduces electricity use at peak times by 2% would avoid costs of nearly 50 cents per kWh, plus reduce the normal cost of consuming the power. Thus, the shift project would have a value of $0.492 + $0.268 = $0.76/kWh, or $760/MWh in avoided costs.

2.5 DISCUSSION

This chapter has explored some of the key aspects of the problems of unsustainabil-ity. The projects of energy transition are not the same as the projects of growth and development of the past century. The primary difference is constraint. Of course, there have always been constraints – we are on a finite planet after all. The transition is where we finally recognize the facts of the carrying capacity of our little space-ship Earth. The alternative energy technologies have not replaced fossil fuels, despite intensive research and development and cost reductions since the energy crisis and environmental awareness of the 1970s. There are unintended negative consequences with many of the new green technologies. We have also examined the belief that economic signals would bring a switch to clean energy and emissions reductions, but the evidence shows this is simply not happening at a scale relevant to the transition to deep decarbonization of the energy system.

The energy transition necessarily drives changes in other areas. Remember that, while there are serious resource constraint issues, the immediate issues are not about running out of oil, coal, gas, land, water or minerals. The immediate issues arise when we ask *how* to change our current land use, buildings, food, products and essentially every other engineered system to use much less energy and achieve down-shift at a profit. This is an interesting situation.

2.5.1 THE BACTERIA IN THE JAR

Let's do a mental experiment. We have a petri dish full of culture medium. That is the nutrient broth, called agar, that biologists use to grow bacteria. We have a type of microorganism that divides every minute. At 9:00 a.m. we place one bacterium in the petri dish. At 9:01 we have 2 bacteria, at 9:02 we have 4, at 9:03 we have 8 and so on. We have done this experiment many times and we know that at 12:00 noon the dish will be full of bacteria, the agar will be gone, and the bacteria population will be crashing as they die in their own excrement. The question is: "When is the dish half full and the food half gone?" The answer: one minute before noon.

This is a simple biology experiment, with a simple mathematical model. But let's make it into a psychology experiment. Let's go inside the bacteria colony and see what the bacteria are thinking. Obviously, during the first 170 minutes the colony is so small it is hard to even see it. Ten minutes before noon, the bacteria occupy 1/1192nd, or 0.08% of the room in the dish. They have had 170 generations of continuous growth. There has been no indication of any kind of problem with food or space or their waste products accumulating to toxic levels.

But let's suppose that at 11:50 the researchers at the Bacter University build a new instrument and discover the walls of the dish. They carefully record their data and do their analysis, and estimate that there is a 60% probability that, given their assumptions, the colony could actually run out of food in 10 minutes, plus or minus 30 seconds. One of the researcher bacteria, Professor B, gives a TEDb talk warning of the limits to growth. The reactions are a bit mixed. It is hard for most bacteria to take the problem seriously as there is nothing in their experience other than growth and unlimited resources. Some are sceptical of the professor and question the science and accuracy of the instrument. Other bacteria point to the fact that the researchers are uncertain whether the so-called limit would be reached at 11:59 or 12:01. The economist bacteria point out that if the food supply gets scarce, then the price will rise, bringing more supply into the market. The superstar entrepreneur bacterium announces that he will build a mega-colony supplied with artificial food. His Bugbook posts get a lot of likes.

Some bacteria call for the government to take action. Others decide to stake out their own patch and defend their food supply. The government pays for more research to more accurately measure the size of the petri dish – but several minutes have elapsed and the dish is now 1/124th, or 0.8% full. The regions of the dish come together to hold a summit, and a historic agreement is made to limit the colony size to less than 2% of the dish, and optimally not more than 1.5%. The growth target is celebrated, but there is no idea how to do it. The government sends some probes out into the lab, and at 11:54, just when the growth target of 1.5% is reached, the government announces the discovery of two new petri dishes full of culture medium.

The economists give TV interviews and explain that the food reserves are just a function of the price and investment in exploration and development, so there is obviously nothing to worry about. Professor B does the calculations and finds that the two new dishes of food would only last 1 minute and 15 seconds, but TEDb doesn't want to book a doomer anymore; people working there only want speakers with positive solutions and uplifting messages.

It soon becomes apparent that the colony does not have a way to move out of their home dish and into another one. The clock ticks past 11:58 and the colony is starting to suffer from the excrement pollution. With 25% of the space consumed, now the whole colony is seeing the limits of the dish. The last two minutes are catastrophic indeed, very much like the end of the world movies that became so popular in the last 10 minutes.

You could use Equation (2.1) or other models for growth to answer the question if you had all the information you needed, like the size of the dish, N_0, and how much space each bacterium occupies, $N(t)$. Let's flip our perspective and look at what is unsustainable about the situation. We can easily see that if the doubling rate is once per minute, and the dish is full at noon, then the dish must be half full at one minute before noon. Skipping ahead and looking at the facts about the future can provide critical insight into the situation. We don't know *how* to change the growth rate – but we have clarity about what will happen if we don't. We don't know what the solutions are, but we can tell that even finding three more dishes full of food would only delay the collapse of the colony until 12:02 if the continuous growth rate continues.

3 Complexity and Communication

Identify what is essential and what is optional.

3.1   ENERGY SYSTEM DATA AND COMMUNICATION

There are numerous sources of energy data available for most countries from government agencies, the International Energy Agency (IEA), British Petroleum (BP) and others. This chapter explores the ways that energy data can be presented and the different ways people interpret that data. We will also look at how visual representations of data can be manipulated to communicate different objectives. You might not know that the idea of graphical visualization of data is less than 150 years old or that the ability to quickly produce a graph of a massive amount of data became available in 1985, when Microsoft Excel was first released for Macintosh computers. Visual representation is a powerful tool in communicating historical trends and in exploring future scenarios. In this chapter, we aim to become familiar with energy data and to learn to be careful about what one can learn from different representations of the data.

3.1.1   ENERGY FLOW DIAGRAMS

We normally think of energy as *flowing* through the energy system, from extraction to conversion, distribution and end-use consumption. Figure 3.1 is a schematic representation of energy conversions, from the primary form found in nature to useful energy services. Primary energy has no commercial value. The primary energy can be in the form of renewable sources, for example, flowing water, geothermal heat, wind, solar insolation and plant biomass. Primary energy is also in the form of finite resources, for example, fossil carbon and uranium. Extraction requires a certain amount of consumer energy for drilling, mining, pumping and building mines or dams. Refining is needed to make different kinds of fuels from extracted crude oil or uranium. Power generation from fuels requires a power plant and cooling water and also produces waste heat. More energy is consumed in moving fuels to customers, and electricity is lost in transmission. These costs are usually included in the conversion of primary energy to energy products. Energy products have commercial value and are purchased by customers by volume or weight of fuel, or kWh of electricity. The energy products do not provide services in themselves. Another conversion process in an engine or appliance is needed to provide the useful energy service. This conversion also produces waste heat. Most of the end-use energy is also waste heat. In particular, older types of lighting, electronics and vehicles have very low conversion efficiency.

The Sankey diagram is a national energy flow diagram produced by the Lawrence Livermore National Laboratory in the United States (https://flowcharts.llnl.gov/

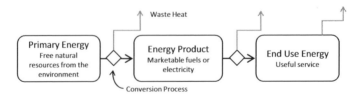

FIGURE 3.1   Basic form of the energy flow diagram.

commodities/energy). The online diagram tool can be used to explore the energy systems of different countries. Find the Sankey diagram for your country and compare to the United States, France, New Zealand and China. You can see that there are large differences in both the primary energy sources and the relative sizes of the end-use sectors. You can also observe the monumental challenge posed by increasing renewable primary energy on a scale to substitute for current fossil fuel consumption.

3.1.2 GRAPHS OF ENERGY DATA

We can understand the dynamic behaviour of systems by analyzing historical data. Analysis means breaking things down into parts and examining individual behaviours and interactions. Figure 3.2 shows the total primary energy production in the United States from 1949 to 2016 (US EIA 2017). All three parts in the figure are exactly the same data but use different chart functions in Excel to represent the data in different ways.

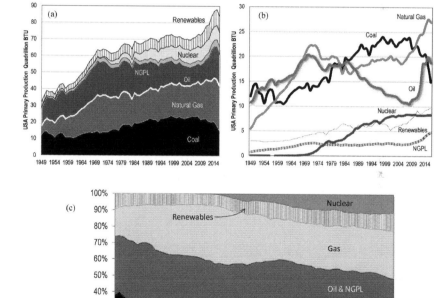

FIGURE 3.2   Primary energy production in the United States by resource in units of quadrillion BTU (quads) where NGPLs refer to natural gas plant liquids; plotted three ways. (a) stacked graph, (b) line graph and (c) stacked percentage graph. (From US Energy Information Administration, Primary Energy Production by Source, 1949–2016, United States.)

Figure 3.2a is a stacked graph, where the historical trend of the bottom-most data set, in this case coal production, is easy to see, but the only other observable trend is the total energy production. Figure 3.2b is a line graph using the same data. The stacked graph shows the total energy consumption as a sum of the different primary sources. The line graph shows the production values from each source plotted together with the same datum. The line graph in Figure 3.2b provides a more positive perception of the contribution of renewable energy, clearly showing the 30% growth in renewable primary energy from 2000 to 2010. When the data is stacked, it is clear that fossil fuel dominates the energy mix, and the growth in renewables is hard to perceive.

Figure 3.2c is a stacked percentage graph using the same data. This stacked percentage graph is commonly used to illustrate energy transitions. Note that the recent steep coal production decline is clear in the common datum plot in Figure 3.2b, but it is hard to perceive in the stacked percentage graph in Figure 3.2c. Conventional US oil production history has a dramatic growth, peak and decline pattern, with the peak in 1974 and then a decline to 1950s production levels by 2010. This classic Hubbert's peak dynamic is clear in the line graph, but it is not evident in the stacked graph. Conventional oil production has continued to decline, and the unprecedented rise in oil production has been from offshore and especially from shale oil extracted with hydraulic fracturing known as fracking. The natural gas and oil boom that started in 2005 and 2006 can be clearly seen in the line graph, but it is hard to see the dramatic 49% increase in gas production from 2005 to 2015 in the stacked volume graph. The stacked volume graph clearly depicts the total volume behaviour.

Figure 3.3 shows the primary production history for only the renewable resources in the United States, using the same data as in Figure 3.2. The line graph shows the development of the large hydroelectricity resource reaching the development limits in the late 1970s. There was an increase in biomass use, largely wood for heating in response to high natural gas prices in the 1970s. Recently, there was rapid growth in solar PV and wind generation since 2005. There has also been a rapid increase in biofuel production, in accordance with the congressional mandates for corn ethanol after 2000. Hydrogeneration has declined slightly, probably due to drought conditions in the western US states, where many large hydropower plants are located.

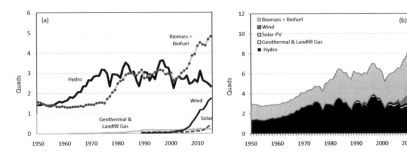

FIGURE 3.3 US primary energy production in quads from renewable resources, 1949–2009: (a) line graph and (b) stacked graph. (From IEA, USA Energy Outlook 2017.)

The stacked graph in Figure 3.3b puts the total wind and solar developments into perspective. When looking at Figure 3.3, do you get any sense of how small the wind and solar primary resources are in the overall mix? In 2015, solar PV surpassed geothermal and landfill gas, but Google Trends reports twice the interest in solar energy topics compared to other renewables. One big problem with communication about energy data is the discrepancy between media attention, public expectation and technical reality.

The energy from coal is clear in all representations, mostly because coal is on the bottom of the stack graphs. Coal energy production has declined by 37% from 2008 to 2016. The usual explanation for production decline is price decline, but the market value of coal has increased 50% from 2000 to 2012, which was not accompanied by increased production. The other reason for decline in production can be depletion of reserves, but this is not the case for coal. Using the Energy Information Administration (EIA) database, we can see that the coal production measured in short tons has also declined, but the average energy content of American coal has declined from 23.1 MJ/kg in 2000 to 22.4 MJ/kg in 2015, which exacerbates the coal energy production decline. The most obvious reason for production decline is demand decline. One trend contributing to coal demand decline is fuel switching of coal to gas in aging, higher-polluting power plants. The biggest factor in the coal production decline is the 50% drop in coal exports since 2012.

3.2   FUTURE ENERGY SCENARIOS AND PATHWAYS

The usual way to construct an energy scenario is to set out assumptions about the future population and energy demand, then to use assumptions about future energy technology costs to determine the energy mix that would satisfy the demand. Projecting historical trends into the future is a common aspect of developing future scenarios. Anticipating the arrival of technologies or resources that have not yet been demonstrated as viable or economical is also a common practice.

Using Equation (2.1) from Chapter 2, we can calculate the average growth rate in oil production of 3.2% from 1950 to 1970. If planners for refineries, pipelines or retail fuel stations used this growth rate to project the future crude oil production to 2015, they would have anticipated 86.9 quads of production. Even with record high prices, which should have stimulated production, the 2015 crude oil production was actually 19.9 quads. Clearly, the crude oil production trend in the 1950s and 1960s was not sustainable, and the growth did not continue.

3.2.1   MODELS BASED ON DEMAND GROWTH

Nearly every national government has an energy-modelling unit dedicated to developing the national energy outlook for future years. Future scenarios are usually mapped out in terms of demand rather than production. Figure 3.4 shows the future outlook for US primary energy production by fuel type from the Energy Information Administration (US EIA 2012). The outlook is for demand to continue to grow and for production to reach 120 quads by 2050. The scenario has modest annual growth of less than 1% for fossil fuels, maintenance of nuclear capacity and faster growth of renewables. Figure 3.4 starts in 1980, and the impression of the scenario is that the recent trends will continue.

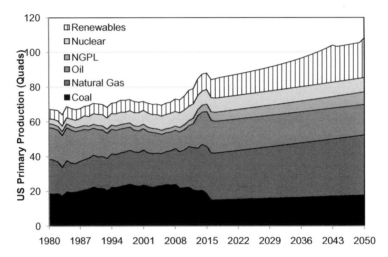

FIGURE 3.4 US Energy end-use projection (quadrillion BTU) by fuel type reflects a scenario for slower economic growth than historical trends. (From EIA Annual Energy Outlook 2012.)

The EIA outlook is for US oil consumption to be reduced by 14% by 2035 largely though much better vehicle efficiency and retirement of the baby boomer generation. But production will grow and the surplus will be exported to China. Coal, nuclear and natural gas demand are expected to remain virtually unchanged, while liquid biofuel and other renewables are seen to provide the expected energy demand growth. This 2012 outlook has the lowest growth rate of future energy use of any scenario the EIA has ever produced. For example, the 2006 Energy Outlook scenario had total energy demand toping 125 quads in 2025 as opposed to the 100 quads in the 2012 scenario. If major infrastructure had been built in anticipation of this high projected future demand, such as upgrading the national grid or building new freeways, then those assets would now be at risk of not producing the anticipated economic benefits.

The EIA projected production of fossil fuels incorporates no response to the issue of atmospheric CO_2 accumulation, partly because it assumes that carbon capture and sequestration (CCS) technology will be developed, even though this would require at least 25% more coal to be used to provide the same electricity generation. The probability is essentially zero that CCS technology will be installed on any coal power plant in the United States by 2025 (Page et al. 2009). According to the International Panel on Climate Change (IPCC) Working Group 3, the decline rate of fossil energy use must be dramatic and sustained in order to manage the risks of catastrophic global climate change (IPCC 2007). According to the report, the most cost-effective approach is improved energy efficiency in buildings and the reduction of coal combustion for power generation. However, the US EIA scenario projects CO_2 emissions will grow continuously at 1.2% annually. Other future scenarios by

International Institute for Applied Systems Analysis (IIASA) and the US EIA also demonstrate a divergence between continued expected demand growth and the necessary fossil energy production decline to mitigate the risks of unsustainable global warming. This tension between continuation of past energy consumption trends and future emission reduction requirements is the driver for the emergence of Transition Engineering.

3.2.2 THE PROBLEMS OF INERTIA AND INNOVATION

Inertia in policy and economics is largely the same concept as mechanical inertia. People resist change. Cultures, governments, businesses, and individuals resist change because it represents a risk. Barring a natural disaster or surprise attack, the activities we did yesterday will be done the same way tomorrow, using the same technology and the same infrastructure and the same energy sources. If that were to change, there is a high probability that it would cause problems. Humans are hard-wired to continue with what they know because they know how it works. We respect and provide powerful positions for leaders who teach and uphold our traditions. Staying with what works is a powerful instinct that has been key to survival. The mechanism for maintaining the system that works is called *tradition*.

The counterfactual dynamic is that people love change. New technologies have provided opportunities for new business growth. Within our population, there are people who are observant and curious and they question the traditional point of view. We idolize innovators, and think of them as historical heroes. One key to human survival has been the ability to change our technology and make better use of resources. We believe that things will get always better because of *innovation*.

Inertia is necessary for survival, but so is change. Technological inertia is mostly a matter of investment. If a technology works, and the factory and market are established, then it is logical to resist change. However, if there is an improvement that can be implemented, or if the old factory needs refurbishment or if the market for the product is declining, then it would make sense to look for changes. There is inherent risk in change, but change also has the possibility of new benefits. The key to survival is always to find the right balance between the risk of trying new things and preserving the benefits of established technologies. It may have occurred to you that, starting with the industrial revolution, we have been in an era of unprecedented change. We are now experimenting, and taking the risks, with new technologies and materials.

A new worldview has emerged during the industrial age. Rather than prosperity being a balance between tradition and innovation, we now believe in perpetual improvement of the human condition and knowledge. We are willing to take huge risks because we believe in *progress*. The point of progress is to continue the project of growth of wealth and expansion of the economy. Even people who are not personally benefiting from wealth creation are convinced that it must continue. Progress is seen as the answer to the problems caused by progress. The simple mathematics of the exponential function should make it clear that material growth cannot continue indefinitely.

3.2.3   The 100% Renewable Scenario

One of the most persistent themes amongst environmental groups and sustainability advocates is the proposition that we can continue the march of progress by substituting clean energy sources for fossil fuels. Companies and cities around the world are setting goals of achieving 100% renewable energy. The United States is the most technologically advanced and resource-rich country. The United States has emitted more greenhouse gas than any other country and thus has a large responsibility for the cumulative warming effect. The United States has the largest installed geothermal electricity capacity and is second only to China in installed hydroelectricity capacity. The United States was the pioneer of concentrating solar power generation and is second only to China in solar PV installations. This renewable electricity capacity, together with the congressionally mandated and subsidized development of corn ethanol conversion capacity, makes the United States a leader in renewable energy development. However, we saw from the data in the previous section that the renewable energy substitution in the United States is objectively small (World Energy Council, 2016). The EIA scenarios are based on future demand, and they do not show any switch to renewable energy.

What would a scenario for the United States switching to 100% renewable energy sources look like? Obviously, the demand for fossil fuels would decline. The 21st meeting of the Coalition of Parties (COP21) reached an international agreement that requires about 6% per year decline in fossil fuel use to reach the target of 80% reduction by 2050. We will use a simple spreadsheet model and a negative growth rate of -6.5% in Equation (2.1) to plot the wind-down of fossil fuel production from about 80 quads today to 16 quads in 2050. The primary assumption for our scenario is the decline of fossil fuel supply, so that is the easy part. Now we need to explore assumptions for the technically realistic development of other energy resources. The nuclear power plants in the United States are to be retired and decommissioned at 60 years of plant life. The original permitted plant life was 30 years, but many plants have been granted operating extensions of 30 years. Coal-fired power plants likewise have a life span of about 50 to 60 years, so we will assume no new construction of coal power plants and the retirement of existing plants when they reach 60 years of age. We assume there is little new hydropower generation built, but that the existing dams and generating facilities are refurbished. The United States added 8,600 MW of wind farms in 2015, so we assume that development continues with 4% annual growth until total wind capacity reaches 20%. A higher wind penetration rate could be possible if large investments are made in the transmission grid, if older turbines undergo an overhaul of their control systems to manage power quality issues, and if turbine manufacturing is greatly increased. We need to consider that by 2030, many wind turbines will be at the end of their life, so the manufacturing capacity will be needed to replace old turbines as well as build new installations.

We will assume that solar PV grows at a high rate of 10% per year in the Sunbelt of the United States, where peak generation matches peak loads. If we assume that panels have a life of 20–30 years, again after 2030, some of the production would need to be used to replace old panels. The experience with the biofuel program indicates that corn ethanol and plant-oil-derived biodiesel will not grow any further. It is likely that the production will be retired in preference for food production. Ethanol requires nearly as much natural gas energy for manufacture as the energy in the ethanol, so our phase-out of fossil fuels will affect the ability to produce ethanol. However, for this scenario, we will maintain biofuel at current levels. Solid biomass for heating is assumed to increase three-fold using high-efficiency combustion systems like pellet fires. Process heat from wastes and geothermal energy is assumed to increase dramatically.

Here are the 100% renewable scenario assumptions:
- Fossil fuels reduce by 6.5% per year.
- Coal declines as 50- to 60-year-old power plants are retired, then maintains 6 quads mostly providing industrial processing energy.
- Oil and natural gas plant liquids (NGPLs) combined production declines from 25 to 1 quad.
- Natural gas production declines quickly from 27 quads to depletion.
- Nuclear power plants are retired after 60 years of generation, and no new plants built.
- No new hydrogeneration as all rivers are already utilized, but hydro remains steady.
- Wind increases at 4% per year and levels off around 8 quads (21% of energy mix).
- Solar PV increases 10% per year to 8 quads (12-fold increase), then replacement of old panels consumes the new manufacturing and generation levels out into the 2040s.
- No further increase in liquid biofuel primary production.
- Biomass energy nearly triples as wood is increasingly used for heating (e.g. wood pellets).
- Geothermal, landfill gas and other biogas from waste increases 10-fold to 2.5 quads.

Figure 3.5 gives the results of our 100% renewable scenario for the United States. The figure is a plot of the percentage of US primary energy production for combined

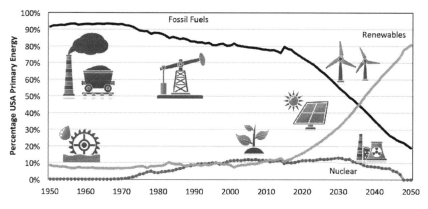

FIGURE 3.5   Scenario results for transition to 100% renewable energy shown as a percentage plot.

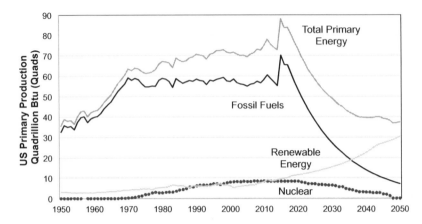

FIGURE 3.6 US primary energy projection to achieve 80% renewable energy supply by 2050: (a) percentage total supply and (b) historical and future primary production values for the 80% renewable scenario.

fossil fuels, renewables and nuclear. Note that the outcome of the scenario provides the results we expected – a reversal of the percentages of primary energy and thus a switch to renewables. This looks great. We made some pretty big assumptions about aggressive renewable energy development, but we see it is possible to transition to renewables.

Figure 3.6 shows the same data plotted in actual primary energy production values. The scenario shows clearly that in order to achieve the target of 80% renewable energy, the total energy production will most certainly decline, likely to the same level as in the 1950s and 1960s. Think about how you feel about the scenario. The idea that energy use could decline into the future does not fit with the popular narrative of green growth, more renewables or a switch to renewables.

Most scenarios assume the need for more energy. Have you ever read an article or heard a politician say that we have enough energy? This is actually one of the core problems of unsustainable energy – reducing energy use is so unthinkable that very few engineers are working on how to do it. The first and necessary requirement for being able to think of innovative and creative ideas is that we must face the actual problems we are trying to solve.

Is reducing energy use to the levels of the 1950s really that unthinkable? The 1950s had a recognizably high quality of life in the United States, but the global population was just over 2.5 billion. The efficiency of energy production, energy conversion and energy use in every sector and every type of technology was much lower than today. There were no vehicle or appliance efficiency standards, and the insulation standards for buildings were much lower and non-existent in some places. Recycling was not done. In the 1950s the average time spent driving to work and school was a small fraction of today. Your grandparents liked to drive their cars. Would they have said that life would be much better, if only they could spend 2 more hours sitting in their cars every day?

> **TRANSITION ENGINEERING CONCEPT 3**
>
> *The fundamental problem for Transition Engineering is the substantial reduction in fossil fuel production and the sustained decline in energy supply and material consumption that will result.*

3.3 CORPORATE RESPONSIBILITY

Companies face challenges in understanding their risk exposure to the climate change, energy and social risks that were discussed in Chapters 1 and 2. Investors and potential partners have an interest as well. Third-party firms have developed a range of methodologies to provide environmental, social and governance (ESG) reports and corporate social responsibility (CSR) rankings and indicators of the corporation's performance in these respects. Most of the ESG rating corporations are less than 20 years old. The methodologies rely on collecting and analysing data that relates to the company's performance with respect to a wide range of factors:

- Climate change exposure
- Carbon emissions, emission trading scheme or carbon tax
- Energy supply, energy intensity, renewable energy
- Natural resources, minerals, supply chain
- Pollution and waste management, ecoefficiency
- Environmental and conservation opportunities
- Social opportunities, employee relations
- Human capital, diversity, human rights
- Stakeholder opposition and community relations
- Product liability, fair trade, abusive labour or environmental liabilities
- Tax strategies and policy influence
- Corporate governance and behaviour
- Customer relationship and brand management
- Preparedness, disclosure and best practice

Socially responsible investment funds have emerged in the market, and they represent conservative and long-term perspectives, particularly retirement funds. Institutional investors have recently become sensitive to the ESG risk factors in their investment strategies. Asset management is the work of monitoring and maintaining things of value. Asset managers may have expertise in engineering or economics of buildings, structures, factories and equipment, but the wider context of ESG is becoming more important in long-term planning. For example, consider a city that owns an airport with runways at sea level. Air travel demand has been growing remarkably over the past three decades, so should plans be made to expand the airport? The price of fuel has more than doubled in 2008, and has risen 30% in 2018. The IPCC report on remaining below 1.5°C global warming indicates dramatic reductions in emissions are urgently required. Should plans be made to reduce flights at the airport? How much risk to the city's income from its airport asset is posed by sea-level rise?

Currently, a number of large consultancies carry out ESG analysis for a company using the company's own data, statistics, government reports and reports from other institutions. For example, Bloomberg ESG Data Service had over 12,200 customers in 2016. Dow Jones Sustainability Index (DJSI) is one of the oldest ESG reports; launched in 1999, it provides a comparative ranking out of 100 for companies across 60 industries. Thus, it does not report on actual sustainability but rather provides a ranking of relative sustainability compared to other companies in the same industry. Institutional Shareholder Services Inc (ISS) calculates a risk-based rating on a scale of 1 to 10, with the first decile indicating relatively low risk over the range of factors. One of the largest ESG companies is MSCI ESG Research, which provides ESG ratings for over 6,000 global companies and also analyzes over 400,000 investment securities in the equity and fixed-income categories. Thomson Reuters ESG Research Data also serves over 6,000 corporations (Huber et al. 2017).

The implication of this widespread interest in analysis, ratings, rankings and monitoring of ESG in the world's corporations is that there must be a market for Transition Engineering. Awareness of the risks of unsustainability is not the problem. A positive approach for Transition Engineering consultants would be to look up the ESG ratings of companies and approach those at high risk. These companies need Transition Engineering work to be done, and they are aware of the areas where they need change.

3.4 POSITIVE APPROACH TO DIFFICULT PROBLEMS

Transition Engineering is the work of developing shift projects that change existing systems in ways that reduce fossil fuel and material use in order to meet greenhouse gas emissions requirements and deal with resource constraints in a profitable and equitable manner. In the previous chapter, we looked at the key problems of unsustainability, including a wide range of pollution issues, waste, loss of biodiversity and industrial agriculture.

Learning about the problems of unsustainable activities can be depressing. In this section, we explore the response to all of these difficult problems of transition. The shift projects achieve energy and material transitions but they step down demand, and that has usually been associated with economic *recession*, *depression*, *contraction* or *collapse*. Thus, we have a lot to deal with psychologically. The problems are serious, complex and global, and there are no solutions, only changes that imply no growth. How do we present these problems, analyse them, process our feelings and those of others, and start finding our way to discovering innovative shift projects? How do we find a positive attitude when the situation is so difficult?

TRANSITION ENGINEERING CONCEPT 4

There are no simple solutions; there are only difficult decisions about what to change next.

3.4.1 CHANGE MANAGEMENT

Change management is a well-known and important specialization in engineering. Change projects are usually done to manage risks, maintain market position, deal with growth or develop new opportunities. All change entails risks, even changes that take advantage of opportunities. Change is often driven by crisis or the failure of some aspect of a system. When a crisis occurs, either the system recovers to the previous state, or it adapts to a new state while preserving essential functions. Change is distressing even when the change project is to improve operations or roll out a new product. Thus, change projects use the language of risk and risk management. Table 3.1 lists definitions of terms we will use frequently in analysis of transition projects.

Risk assessments analyse the probability of the occurrence and the likely impacts of a particular issue within a certain timeframe. Risk management is the process of building defences, adding resilience and developing adaptive strategies to deal with issues before they become problems and cause a crisis (HM Treasury 2004). Risk management is most well known in relation to preparations and fortifications against natural disasters. Most risk management strategies were developed after disasters exposed vulnerabilities. For example, many of the functions of the weather service are to give people warnings and advice on how to avoid risks in extreme weather events.

Transition Engineering of transport, electricity and heating systems may cause distress because people rely on these systems for essential services, and it is true that if there were suddenly no fuel, the transport system would break down in a catastrophic way. However, shift projects are about improving energy adaptability and managing long-term energy supply risks, while eliminating the use of fossil fuels through changes in the way the system works. Most people would see 'running out of oil' as a risk. But a community, a school, a manufacturer or any organization that redevelops their operations to avoid relying on fossil fuel would then have the freedom to operate into the future with much more resilience and mitigated risk.

TABLE 3.1

Definition of Key Terms Used in Change Management and Transition Engineering

Issues	Important potential problems that have negative effects if they occur
Risks	Exposure to danger or loss due to an issue that could occur over a given period
Transition	Planned change from current high-risk state to a low-risk state over time
Resilience	Ability to recover from a crisis or disruption
Adaptive capacity	Ability to change while maintaining essential functions
Adaptive strategy	Planned changes to deal with different pressures and problems
Scenario	Science-based model of changes in energy supply or end use over time
Idea	Potential change, untested for effectiveness, benefits or cost viability
Opportunity	Feasible, effective and viable potential change in a current system

3.4.2 Wicked Problems of Unsustainable Energy

Wicked describes a problem that defies solution. Wicked problems are so difficult that even getting started can be impossible. Wicked problems are inherently complex and are often global and long-term. Transition Engineering deals with wicked problems. We usually think about problems because of the harm they cause. Normal problems have solutions that manage the harm while preserving the benefits. The catalytic converter on automobiles is a solution to the problem of vehicle pollution. The wicked problem is that automobiles work great. Everyone has an essential need for access to work and activities, goods and services. But oil use in the existing vehicle fleet is not sustainable. The personal vehicle traffic in every city is congested, and the infrastructure is the biggest public expenditure. Even a small fuel supply disruption would throw the city into chaos, but the carbon emissions must decline rapidly. Cycling and public transport are more sustainable options, but the cars make cycling too dangerous for most people, and it is impossible to serve more than a small percentage of the population with public transport in an urban form dominated by personal vehicles.

3.4.3 Responding to the Problems of Unsustainable Energy

The problems of global warming and energy and resource scarcity are serious and present serious risks. The problems have been discussed in scientific and engineering terms. Movies and the popular media play out our sense of a bleak future where resource scarcity, climate change or some other humanmade disaster will cause the collapse of civilization and civility. As we conclude this chapter, we will understand the psychological response to wicked problems and frame our own response so that we can manage it and get on with the work of engineering the transition.

We fear change. We fear not knowing what to do. We fear losing something valuable. Fear does not improve creativity and problem-solving capability. The problems of unsustainability are serious, and the science should make everyone afraid of the consequences. Recognize that you and your colleagues, and the population in general, experience well-known psychological responses to the mega-issues of global warming and energy reduction and the other wicked problems.

There are stages we go through in order to deal with the facts of the problem and emerge mentally prepared to be creative and innovative. Be aware that we wish to skip over the stages. We want to get rid of the fear as soon as possible, and so 'positive solutions' to the problem are appealing. Positive attitude is important but not sufficient for engineering projects. Let's say that you decided you wanted to climb Mount Everest. A positive attitude would be important, but if you weren't prepared, didn't have the right guide and didn't know the right route, then positive solutions would not be of any use.

Figure 3.7 shows the stages that all people progress through as they learn about the problems of unsustainable energy and climate change. The problems are so dangerous that shock and denial are natural reactions. The main reason for the disbelief reaction is that there is no solution to the problem. People who get stuck at the disbelief stage can find plenty of websites and articles that support their position. Once you move past the shock, the facts of the problem become undeniable. The second

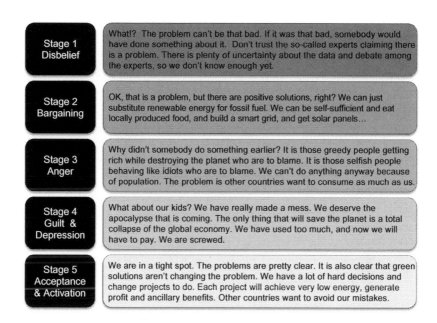

Stage 1 Disbelief	What!? The problem can't be that bad. If it was that bad, somebody would have done something about it. Don't trust the so-called experts claiming there is a problem. There is plenty of uncertainty about the data and debate among the experts, so we don't know enough yet.
Stage 2 Bargaining	OK, that is a problem, but there are positive solutions, right? We can just substitute renewable energy for fossil fuel. We can be self-sufficient and eat locally produced food, and build a smart grid, and get solar panels…
Stage 3 Anger	Why didn't somebody do something earlier? It is those greedy people getting rich while destroying the planet who are to blame. It is those selfish people behaving like idiots who are to blame. We can't do anything anyway because of population. The problem is other countries want to consume as much as us.
Stage 4 Guilt & Depression	What about our kids? We have really made a mess. We deserve the apocalypse that is coming. The only thing that will save the planet is a total collapse of the global economy. We have used too much, and now we will have to pay. We are screwed.
Stage 5 Acceptance & Activation	We are in a tight spot. The problems are pretty clear. It is also clear that green solutions aren't changing the problem. We have a lot of hard decisions and change projects to do. Each project will achieve very low energy, generate profit and ancillary benefits. Other countries want to avoid our mistakes.

FIGURE 3.7 The stages of realization of the problems of unsustainable energy and material use and the accumulating environmental degradation.

stage of bargaining is an active seeking of positive solutions. Again, it is understandable that, when you understand the problems, you would want to find solutions. But if you get stuck at this stage, you will become fixated on things like substituting fossil fuel with renewables, which objectively will not solve the problem. When you move on to stage 3, you may get angry at the hubris of people and inaction of leaders. Study the mistakes of the past, but don't get stuck in anger. Move on and go through the grieving process for the consequences already locked in, the opportunities already lost and the damage already done. But then turn your focus to digging into the data and modelling of the real situation in a particular place and for a specific essential activity. Move on to the final stage of acceptance and activation. Let the fear, anger and grief transform into resolve. Seek out a clear focus for action – changing the unsustainable systems.

Take a survey of your own attitudes, and discuss with others where you are in the process. If you are at the point of fully understanding the problems of unsustainability, then you are ready to get to work on Transition Engineering. Recognize that others may be stuck at different stages. Don't spend time trying to convince them that their response is wrong, because each of these stages is actually part of the natural response. Rather, think of ways to help other people progress to the next stage from wherever they are. It is difficult to form an effective Transition Engineering team unless all members have progressed to the acceptance and action stage.

People tend to form into groups that share their perspective. Transition Engineering will inevitably involve people with different perspectives about sustainable development. Table 3.2 presents a snapshot of the perspectives on the problems and the

TABLE 3.2

Transition Innovation Brainstorming Process Requires That Participants Acknowledge Existing Perceptions of the Nature of the Wicked Problem and Solutions and Learn to Recognize Creativity-Defeating Beliefs in Themselves and Team Members

Perspective	Problem	Solution
Doomsday	A religious or supernatural power has ordained the end of the Earth, or an extraterrestrial event is inevitable and beyond our control.	Follow religious teachings and wait for the end times or for the comet to strike the Earth. Alternatively, plan on moving to Mars.
Techno apocalypse	Failure of complex infrastructure and energy systems will lead to the disintegration of civilization.	Become self-sufficient, learn survival skills and prepare to defend your family and your supplies.
Environment-induced collapse	Like many civilizations before us, we have become too complex and done too much environmental damage; thus our food supply will fail, and civilization will collapse.	Make political leaders aware of the environmental damage and call for action.
Limits to growth	Growing and continued use of finite resources in the past 70 years has led to overshoot of carrying capacity in population.	Return to past levels of population and resource use, relocalization of food supply and stop the growth of product and material consumption.
Green energy technology substitution	Belief in gross domestic product (GDP) and continuous growth by economists has led to development of 'dirty' energy sources and suppression of 'clean' energy solutions.	Substitute fossil fuels with renewable energy sources through competition, and application of subsidies and incentives.
Factor four	There is no real problem with current economic models; the problem is that technology needs to develop faster by developing our natural capital of human ingenuity.	Technology advances in efficiency and use of new materials, new sensors and automation will provide continued economic growth, and the market will supply our energy needs.
Invisible hand	Market signals are not valuing ecosystem services sufficiently to create demand for more environmentally friendly technologies.	Put a cost on externalities, polluter pays, carbon market, carbon tax, incentives for renewables; fund research for alternative technologies.
Social justice and sustainable development	Morality and values for equity and end of poverty must evolve through political struggle to change social norms. The problem is that the well-being of the environment, third world and future people has not yet been recognized as a current responsibility of first world people.	Influence current values through social consciousness movements; put pressure on corporations and governments to reflect these sustainability values; bring court cases defending rights of environment, animals and future people. Provide development aide for impoverished communities.
Transition towns	Peak oil, climate change, economic instability	Organize community-level projects for local self-reliance and energy independence, for example, permaculture gardening, farmer's markets.

solutions espoused by these groups. Once the group forms an identity and gains a following, the members often become dedicated to a particular solution to the problems they perceive. All of these perspectives have valid evidence and a valid rationale, and the changes they are seeking are important. However, we should look at the evidence that social movements and economic instruments have not changed the Business-as-Usual (BAU) trajectory of industry, technology, manufacturing, extraction, infrastructure and energy systems, which are, by design, causing the problems.

3.4.4 History of Transition Engineering

Engineered systems provide essential services, infrastructure and production both for profit and for the common good. However, successful technologies have almost always had unintended consequences. Scientists measure the damage that new technologies cause to human health and the environment. Eventually changes are made to reduce the pollution and health issues caused by new technology. Table 3.3 gives some of the failure events over the past century that led to the emergence of new engineering fields as discussed earlier in this chapter. The history of engineering innovation is that it leads to new benefits and economic growth; then the problems it causes are recognized through catastrophic failure, and the engineering profession emerges to change the way systems are designed and managed.

Innovative engineers recognize a risk and are inspired to develop new, safer products. These new products sometimes take some time to be adopted. Once the products are brought to the market, they are required by regulation, and the older versions are phased out. Safety or environmental regulations are enacted *after* the engineering solutions are available. For example, engineering technician J. W. Hetrick was

TABLE 3.3

The Emergence of New Fields of Engineering That Developed New Practices and Technologies in Response to Problems or Disasters

Engineering Field	Indicative Problem or Disaster	Timeline
Water and sanitation	Broad Street cholera outbreak, London	1854
Safety	Triangle Shirtwaist Factory disaster, New York	1911
Natural hazards	Great San Francisco Earthquake, California	1906
Emergency management	Massive earthquakes and hurricanes, United States	1962–1972
Waste management	Pollution seepage from dumps in most cities	1965
Environmental	Cuyahoga River, Cleveland, and Lake Erie declared dead	1969
Toxic waste management	Love Canal disaster	1978
Reliability	Communications, aerospace, military, power grids	1960s
Energy management	Organization of Petroleum Exporting Countries (OPEC) oil embargo and energy crisis	1974
Renewable energy	1970s oil crisis, Kyoto Protocol, Rio Earth Summit	1997
Transition	Climate change, peak oil, ethical responsibility to future generations	2015

inspired to invent the automobile airbag after having an automobile accident in 1952. The first car sold with an airbag was in 1973; by 1990, air bags were standard in many vehicles, and in 1998 the US federal government mandated dual front airbags on all passenger cars (McCormick 2006).

The first tenet of Safety Engineering is to be honest about failures and possible remedies, and to bring issues to the attention of co-workers, employers and the public (ASSE 2017). Safety Engineering grew very rapidly because the changes developed through innovation and research were immediately deployable solutions that reduced risks for workers *and* factory owners. Adopting safety protocols and technologies always means changing the existing system and behaviours.

Every industrial chemical that has been discovered to cause cancer was developed to provide a benefit and was profitable. When the detrimental effects are discovered, the chemical companies always seem to resist changing or banning the product to reduce the hazards. But then somewhere in the industry, some engineers started thinking about how to change the chemistry. Companies that are leaders in adopting changes and working with regulators to develop standards are often in a much better competitive position than those that resist change. For example, Apple has led the industry in both profits and elimination of toxic chemicals (*n*-hexane and benzene) and toxic materials (polyvinylchloride and brominated flame retardants). Companies that develop new systems early will be able to transition their products to the new platforms before the cost of the old materials soar in response to manufacturers of banned substances shuting down production. For example, refrigeration systems worldwide have been redesigned because of the phase-out of ozone-depleting refrigerants, and they are currently undergoing redevelopment to use working fluids with lower greenhouse gas potential (Linden 2014). Companies that hang onto their old designs for too long will have to pay escalating prices for the old refrigerants as the suppliers shut down production.

3.5   DISCUSSION

In this chapter, we have explored the importance of perspective. Perspective can even influence how people understand objective facts. Engineers are applied scientists. Engineering uses physical science to derive understanding of phenomena by objective observation and testing of hypotheses to develop repeatable models. The engineer's job is to use objective facts and reliable models to 'make things work'. It would be fair to say that most engineering work is relatively disconnected from the behavioural sciences. Engineering work is technically challenging but psychologically uncomplicated. Society relies on industrial activities, and it is up to engineers to practise social responsibility. When beneficial engineered systems and products are causing harm, cognitive dissonance can make things seem difficult and complicated. Cognitive dissonance describes the uncomfortable mental state one experiences when simultaneously holding contradictory beliefs, ideas, values or perceptions. Psychologists say that cognitive dissonance is so uncomfortable that people will not remain conflicted for long and they will choose sides. Although they know both things are true, they will pick one and discount the other.

FIGURE 3.8 Cognitive dissonance means simultaneously holding contradictory ideas. In this image, it is difficult to see two contradictory things at the same time. An old person and a young person are both there; you must control your perspective in order to flip between them.

How can you hold conflicting ideas in your mind at the same time? Optical illusions like the ones in Figure 3.8 are fun to look at because of the difficulty our minds have in seeing two things at once. Each drawing can be perceived two ways: as a young person or as an old person. Can you see both? Can you control your perception so that you can see the old person when you want to and the young person when you want to? Is it hard to admit that both images are there simultaneously? Of course, we see these drawings as 'tricks of the eye'. But they are good illustrations of the situation when we are at a transition.

- Oil is the best fuel ever developed. Oil supply and use is the biggest risk to a sustainable society.
- Fossil fuel has allowed us to develop to the most advanced civilization of all time. Fossil fuel is the greatest risk to the future of civilization.
- Our exploitation of minerals and natural resources has provided economic growth and a higher quality of life. Our exploitation of minerals and natural resources are causing catastrophic biodiversity and habitat loss and toxic pollution.

The situation during the transition era is exemplified by this situation of holding similar yet opposite and counterfactual ideas at once. We can imagine that this was the situation that the first 62 industrial engineers were in when they got together in 1911 after the Triangle Shirtwaist Factory disaster. The industrial revolution had delivered progress and wealth. Factory manufacturing made goods affordable for more people. But workers were dying in factories in ever-greater numbers. The engineers could not shut down the factories in order to save the people; they needed to work in a new way. Their idea to *prevent what is preventable* and to be honest about the benefits and risks, problems and solutions was the real foundation for what we now call Transition Engineering.

In the remaining chapters of this book, we use the Transition Engineering tools to help us discover the down-shift perspective while still being able to understand the growth perspective we normally see. In the down-shift era, *not* using energy and materials is the preferred option that has benefits we can discover and develop when we shift our perspective.

3.5.1 The Fox and the Hedgehog

The fox knows many things, but the hedgehog knows one thing.

Greek Poet Archilochus 680–645 BC

One perspective on decision-making is that of the fox. The fox is a linear problem solver. The fox is a predator. The fox's survival depends on being generally successful in spotting any kind of opportunity for hunting prey. The fox is working toward an immediate objective of getting a meal. There are any number of options that might arise. The fox will have to work quickly to recognize potential, use existing capabilities and pounce on opportunities. The fox may have to think quickly to make the most of each opportunity, using past experience and quick problem solving as the prey takes evasive actions. The consequences of taking risks are low. The strategy for survival is to try many different things and be successful on just a fraction of attempts.

Another perspective on decision-making is that of the hedgehog. The hedgehog is a systems thinker. The hedgehog is prey. The hedgehog's survival depends on not getting eaten. The hedgehog is working toward the goal by being aware of danger and taking precautions. The hedgehog is successful in its objective every day that failure does not happen. Taking risks increases the probability of failure. The hedgehog's approach to the problem has a long-term perspective, and his strategy is to be alert and repeat what works, like tucking into a ball and presenting only its sharp spines when it is alarmed. The hedgehog also has a strategy to find a place to hide under a log or rock. The hedgehog is thinking about the big picture. The strategy for survival is not to fail, not even once.

The fox is an opportunist, and the hedgehog plays it safe, but what if there is a bigger problem?

The comparison between the fox and the hedgehog reflects the two basic ways of dealing with information and approaching big decisions. The story does not say that one perspective is better than the other. Humans have been a successful species so far, and societies have always had people with both perspectives. Some people take risks and try a lot of things, and some people are cautious and do what has always worked. In a time of surplus, when survival is not an issue, risk-taking behaviour can return rewards without too much concern for consequences. In a time of crisis, a short-term perspective is needed to respond to emergencies. The long-term, big picture, precautionary perspective has also been essential for preserving traditions and practices that are sustainable.

Transition presents a new dilemma for both kinds of thinking. Consider the hedgehog who knows he can't outrun a fox, so he runs onto the railroad tracks and tucks himself down between railroad ties, not knowing that a train is coming. The fox is working on the short-term problem of how to dig the hedgehog out from his hiding place, without getting poked by the hedgehog's spines. Both have lost sight of the big picture. The fox forgets about the whole system context in which he is trying to solve an immediate problem of getting a snack. The hedgehog forgets about the need for change when he is trying to solve an immediate problem with the same strategy that has always worked before. Will the fox and the hedgehog become aware of the bigger problem coming at them and both refocus their attention and take new kinds of action? After they deal with the bigger problem of the approaching train, they can always go back to the predator-prey daily routine.

We need quick thinking to manage immediate and obvious risks. We need long-term thinking with dedication to doing what has worked in the past. But at a transition point, we need both perspectives at the same time. We also need to look up, take immediate action and not get paralyzed by fear.

Energy transition is a wicked problem of survival. We know that the problem is unsustainability, but we risk inaction and ultimately catastrophic failure if we cling unwaveringly to the hope that the sustainable society will somehow happen using the positive solutions we have already been trying: virtuous consumer choices or alternative technologies that maintain growth in consumption of renewables. We also risk catastrophic failure if we keep being distracted by miracle technologies and empty promises. Transition Engineering requires both perspectives. We need our 'right brain' abilities to keep the big picture in view, while we also need our 'left brain' abilities to focus on immediate projects to provide for essential needs. We will cultivate the creativity and flexibility to pursue many opportunities, while also being relentlessly sceptical and realistic about possible options. We understand the power of hope while working to discover viable prosperity through down-shift and sobriety.

4 Transition Engineering

Draw a picture.

The mission for Transition Engineering is the redesign and redevelopment of existing systems. The objectives are to down-shift to ultra-low energy, mineral and natural resource consumption while uplifting quality of life, environment and real value. The ultra-low consumption system will manage the main issues of unsustainability and will support the activity systems of the population. The Transition Engineering approach is aimed at discovering and carrying out *shift projects* that change an existing system to much lower consumption of energy and materials. Nearly all of these shift projects will require innovation, and the possible innovations are not limited to technologies. Energy transition innovations require new perspectives; creativity in business and economics; strategic ingenuity in policy; and novel understanding of human, social and environmental well-being.

We are at a point of transition, where the future is not like the past, everything is going to change, and we don't know how to accomplish these kinds of changes in productive ways. The solution that has a reasonable chance of working is to stop doing what doesn't work. An essential rule of engineering is *draw a picture* because it helps to understand complex concepts, maps out the parts of the project and describes the sequence of the workflow. Figure 4.1 is a picture embodying the approach we will study in this chapter.

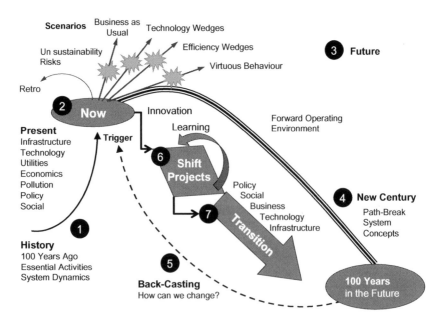

FIGURE 4.1   Interdisciplinary transition innovation, management and engineering (InTIME) approach framework.

4.1 DEFINING THE SYSTEM AND THE InTIME APPROACH

Figure 4.1 shows the seven-step approach used to discover down-shift projects. It is called the interdisciplinary transition innovation, engineering and management (InTIME) approach. The preliminary step for an InTIME project is to clearly define a particular activity system involving a group of people in a specific place. An example of a transport activity system definition would be the travel to primary school by children in Santa Fe, New Mexico. An example of an organizational activity system would be conference travel by academics at the University of Bristol. An example of an industrial activity system would be the milk-processing cooperative, Fonterra, in New Zealand.

New ideas require new perspectives, and the InTIME approach is designed to help you 'flip your perspective'. Each of the seven steps contributes to understanding complex systems in new ways and thus provides new perspectives (Krumdieck 2013). Once a system has been defined, and the problems of unsustainability are understood, then we get to work to discover a productive and effective down-shift project InTIME.

Step 1 Study History. Take a step back and do some research on the history of the activity system. This is the first step because it makes sense to learn about the past and to examine how the system developed to the current point. It is also the most easily achievable step. Given that wicked problems are impossible, a good first step is to take on a project that is both interesting and relatively straightforward. The outcome is an understanding of the way people carried out the essential activity 100 years ago.

Step 2 Take Stock. Gather data about the current situation in the specific location and for the specific activity system. Explore the interconnected systems and the social, political and economic context. The outcome is a characterization of a specific current wicked problem.

Step 3 Explore the Future. Develop and explore future scenarios. Projecting past trends into the future is the usual approach, and we have a few interesting new ways to look at scenarios. Then we will explore the forward operating environment (FOE), which simply means applying the requirements for fossil fuel reduction and any other constraint to the activity system out to 2100. The outcome is a clear picture of what is not likely to happen due to a few key fundamental constraints.

Step 4 Time Travel. Creative brainstorm this place and this activity system 100 years from now and ask, "How are they carrying out these essential activities?" The outcome is a believable picture of people like us living in this place and not using unsustainable energy and materials.

Step 5 Backcast and Trigger. Take the lessons learned from the future and the past, and back-cast to the present, addressing the key questions: What is actually different? If it is possible that there is a future where our wicked problems have been worked out, then what was the trigger that happened now that leads to that future? The outcome is a clear picture of the types of triggers that would cause a change in business as usual.

Step 6 Down-Shift Project. Brainstorm and explore to discover concepts for different kinds of shift projects that could deliver feasible changes now, and that would still be providing benefits 100 years from now. Build a learning cycle into the shift project plan. The shift project must be feasible and viable, mitigate risks, provide benefits, and lead to a break with the current business as usual. The outcome is an innovative shift project idea and project plan.

Step 7 System Transition. Investigate transition management scenarios for where the new direction leads. Develop a story of how the trigger and the shift project lead to adoption and uptake and more projects in other systems, and how that leads to a new energy transition pathway. The objective is to discover how the shift in direction leads to growth in new businesses, social benefits and environmental regeneration.

4.1.1 Defining Activity Systems

Figure 4.2 shows the activities that human societies undertake in daily, weekly and monthly patterns. This web of life has had as many expressions as there have been tribes, cultures or civilizations in every corner of the world, with an inspiring array of ways to live in vastly different environments with different resources. All people in all times and places have used energy and materials. Energy and

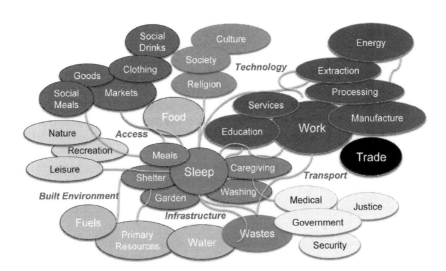

FIGURE 4.2 The essential activity system of a society is essentially the same in every human settlement throughout time on every continent, even though the built environment, infrastructure, technology and specific behaviours are different. Activity centres have always been strategically located to provide access to the activities. (Reprinted by permission from Springer Nature: *Transition Engineering* by Krumdieck 2017.)

material use is a consequence of human activity. *Activity systems* are the regular pattern of behaviour of a person, family, group or community. For example, people have daily activities of farming, building, manufacturing, providing services or attending school. These activities are 'work' and 'education', which also define the purpose of the built environment, the buildings where the activities happen. The transport activities are the trips between the origin (home) and destinations. Economic activity systems involve trade or exchange of money for goods or services. Essential activities are crucial for survival. Some activities are necessary for quality of life, social well-being and culture. Optional activities support lifestyle, preference and convenience. The key engineering principle relating to activity systems is that energy and material use are determined by the technology and built environment that already exist, and if the technology changes, the activity system adapts quickly.

The key to human survival may well be the capacity to adapt to local ecosystem and resource availability through evolution of culture, technologies and traditions. There are two dynamics to this adaptive capacity: resistance to change to manage risks, and rapid change to take advantage of opportunities. When a way of life works sustainably, it is embodied in culture and tradition. The young are educated in the 'way we do things' and brought up in the reinforcing culture. Language, religious practices, festivals, cuisine and architecture styles are examples of traditions that resist change and thus help to preserve sustainable aspects of a society. The other adaptation mechanism is the small percentage of people who are change makers, problem solvers, disruptors and innovators. The technology and tools that we take for granted were each invented by one person. Archaeologists can find the point and time of origin of game-changing innovations like pottery making, iron smelting, writing or saddle making. Historical evidence shows that most innovations happen as singularities but then spread rapidly throughout networks of people with related cultures and trade relationships. Thus, the essential activity system remains the same, but the ways the activities are carried out are simultaneously static and fluid, resilient and adaptable. The ability of the energy system to provide for essential activities and quality of life is the important measure of the prosperity of the community. Thus, when we analyze a particular energy system to determine if there are problems of unsustainability, we will focus on the risks to essential activities.

A fundamental assumption of economics is that people act rationally and their actions are driven by the motivation to increase their wealth and consumption. The use of the term *economy* is relatively modern, developed during the Great Depression to refer to the collective primary production, manufacturing and labour. The most common measure of the economy is the total money spent, called gross domestic product (GDP). GDP is a very rough measure of social benefit, because it doesn't distinguish between money we spent for things we wanted, things we needed or purchases we were forced to make because of a disaster. The fundamental assumption for transition innovation is that the economy works if the activity system works. Every society throughout time and in every place has had an economy. Given even the most challenging conditions, people have always found ways to carry out their activities to meet their essential needs, and they have always had economic transactions. The point is that consumption is not the reason for society's activities; rather, the society's activities are carried out to meet essential needs, and they involve economic transactions.

4.1.2 Wicked Problems of Activity Systems

By definition, for innovation to be possible, there must be no currently known solution. If there was something that worked, then you could just use it, and innovation would not be needed. Wicked problems involve activities that are beneficial and profitable, but simultaneously harmful and unsustainable, as illustrated in Figure 4.3. The energy transition for nearly any activity system poses a suitably wicked problem to provide fertile ground for innovation. The first rule is to set the system boundaries and define the activity system. The second rule is to draw the picture describing the wicked problem. Do not attempt to identify a solution at this point.

A wicked problem is not pleasant to contemplate and is likely to cause cognitive dissonance, the mental discomfort of simultaneously holding conflicting motives, ideas or beliefs. It is helpful to name the wicked problem and write down the conflicting truths about it using the diagram in Figure 4.3. Wicked problems do not have a one-dimensional solution. Wicked problems are also resistant to a standard linear engineering approach because of the circular nature of the positive and negative aspects, and the different timescales of risks and benefits. We use the seven-step transition innovation methodology for the same reason we use methodologies in all engineering disciplines; because without the methodologies, nearly all kinds of engineering problems would be unapproachable. We advise engineers to work through the approach with a team including people with a range of expertise in the activity system, built environment and technology.

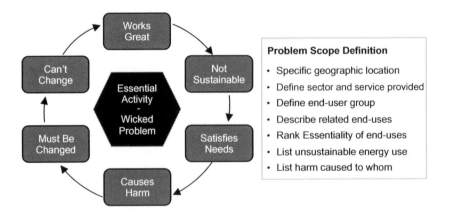

FIGURE 4.3 A wicked problem is an essential activity system that simultaneously functions but is unsustainable, provides benefits and causes harm and must change but resists change.

Example 4.1

Primary school children in Santa Fe, New Mexico, travel from their house to school. Education is the essential activity, and the place is the neighbourhood surrounding each school. Children must get to primary school safely and on time. Parents drive their children in order to make sure they are safe. The cars work great: they are comfortable, safe and affordable. The main safety hazard for children walking or cycling to the school is the heavy car traffic because of all the children being dropped off. Children being driven in cars to school is the problem. The city spends a lot of money on policing of school zones and building traffic control, drop-off and parking facilities. The school drop-off trip is one of the major sources of traffic congestion in the city. The use of cars is affordable, but there are high external costs. About 12% of the car trips in the city are for the purpose of primary education. Although the fatality rate has declined 56% since 2000, motor vehicle crashes remain the number one cause of unintentional death among children 18 years and younger. All of the vehicles used by parents use fossil fuel, and students are exposed to emissions because of the high concentration of vehicles idling and the engines not being warmed up because the trip is the first of the day and is less than 10 minutes in duration. This activity system is unsustainable, so it must change. Conversely, parents could be considered neglectful if they allowed their children to walk or cycle through the congested area to school, so they do not want to change.

4.1.3 InTIME Brainstorming Process

Brainstorming is a well-known exercise in product development and creative problem solving. A series of dedicated brainstorming sessions are usually needed where the participants focus their attention on exploring the system through the first five steps before discovering shift project ideas. A brainstorming session has a particular topic and objective. Brainstorming teams use stories and tricks that break from traditional mindsets and perceptions. For example, all participants are encouraged to toss in ideas, and all are encouraged to think outside the box and to propose things that may even seem impossible. A standard rule for brainstorming sessions is that all ideas are recorded and that criticism and critique are not allowed until after all creative contributions have been exhausted. It is widely recognized that fun and laughter are essential to brainstorming creativity. A brainstorming session usually requires a moderator to ensure the brainstorming rules are followed. Brainstorming methods minimize participant self-censorship and elicit creativity. Interpersonal psychology can derail workshop activities; usually outgoing personalities dominate discussions and rules must be set out and followed to ensure full participation (Hovatter 2013). Separate brainstorming workshops are used to work through the different aspects of the InTIME approach:

- Management – Steps 1–3: Study history. Gather the data, research and modelling and expertise of different stakeholders. Work to understand the relationships between technology and energy, quality of life and activity systems. Explore future scenarios, with careful attention to technical viability and changing relationships.

- Innovation – Steps 3–5: Flip your perspective. Question all assumptions, and construct a probability map for one or two things that we do know about the future (normally 80% less fossil fuel). Explore the activity system 100 years in the future as if reconnoitring for a *Lonely Planet* travel guide. Cast back to the present, with comparison to the past, and brainstorm the triggers for transition – what starts now that leads to the 100-year concept?
- Engineering – Steps 5–7: Brainstorm possible triggers for change that could also provide new opportunities. Brainstorm shift project concepts and carry out strategic analysis to identify the opportunity space. Evaluate the transition pathway that would follow from the learning around the shift project and expansion and uptake from the disruptive innovations.

Energy transition teams need to have diversity, but people with different disciplines and perspectives typically find effective communication particularly challenging. Different professions have their own language and their own interpretation of common terms that might have different meanings. For example, the word *efficiency* can be used to gauge the first and second laws of thermodynamic energy conversions in power generation, it can mean the relative heat transfer or thermal resistance in building design, and it means maximizing profit in economics. In addition, the English language uses the same words to mean different things. For example, the word *design* is used in many different parts of the creative and engineering professions where the actual purpose and methods used are quite different. The main point is that the InTIME method requires coach for the brainstorming methods to design and run the transition brainstorming sessions (Krumdieck 2012).

4.2   STEP 1: STUDY HISTORY

The first step involves research into the historical data about the specific system of interest and the historical context. The objective is to understand the system dynamics by observing past behaviour. How were things 100 years ago, and how have they changed over time? What are the resources that have always been important, and what are the things people have considered important at different times? The main reason to study history as the first step is that wicked problems are so hopelessly unsolvable that even having discussions is difficult, and the creative work of brainstorming is stifled. Most people are not knowledgeable about the past 100 years of historical development of a particular technology. In energy systems, this means looking at historical energy supply, end-use technology and consumer or operator behaviour. Step 1 aims to characterize the effects of different technologies, resource developments, policies or social changes on energy and resource demand of the organization or system under study.

Explore local history, focusing on local culture, drivers for change and examples of problem solving. This step is fun and a great ice-breaker for engineers and stakeholders in the transition team. Creativity cannot occur in a humourless vacuum, and problem solving is crushed by negativity. Do not proceed to the next step until the energy transition team has a solid understanding of the fact that, 100 years ago, people thought just like we do today, and they did not think about us at all. Do not try to formulate solution ideas for current wicked problems by going back to the past. The people in the past made their decisions in their own context based on their knowledge at the time. We are seeking only to learn the history of the activity system, how it worked under different circumstances and what caused change in the past.

4.2.1 Modelling System Dynamics

Three typical growth dynamics are illustrated in Figure 4.4: exponential, logistic and linear. The initial growth is slow for all three models, and it is hard to distinguish one type of growth from another. However, the logistic and exponential growth appear to suddenly rise rapidly, with the exponential growth surging to much higher values than linear or logistic growth. In natural systems, exponential growth is common for uncontrolled growth, like single-cell algae and bacteria blooms. Logistic growth models the recovery of species in a forest after a fire. Linear growth models the accumulation of sediments or the annual mass increase in a tree. As we have already explored in Chapter 1, exponential growth is actually the first half of a boom-and-bust cycle for most human activities, while the logistic model can be used for technology uptake, market penetration and change to a sustainable level of activity. The historical population, consumption and CO_2 growth patterns since 1950 can be modelled by the exponential function. Once the historical data has been ok to one or more of the growth models, it can be used to explore possible scenarios if the historical patterns were to continue.

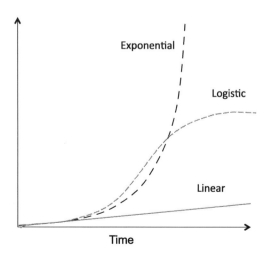

FIGURE 4.4   Three types of growth dynamics for growth in a quantity with time $N(t)$.

The future value of a quantity, $N(t)$, growing at a constant rate is given by:

$$N(t) = N_0(1+r)^t \quad \text{exponential growth} \tag{4.1}$$

where N_0 is the initial value; r is the growth rate per unit time, t.

Linear growth occurs if the same amount is added to the previous year's consumption, given by:

$$N(t) = N_0 + t \cdot N_c \quad \text{linear growth} \tag{4.2}$$

where N_c is the constant amount added each year and t is the number of years.

Logistic growth is experienced when growth from the initial value is exponential at the beginning but then reaches an inflection point, the growth rate slows and the long-run value approaches a new value, N_B:

$$N(t) = \frac{N_B}{1 + \left(\dfrac{N_B}{N_0} - 1\right)e^{-kt}} \quad \text{logistic growth} \tag{4.3}$$

where k is the absolute growth rate. The S-shaped logistic growth curve is often used to model the market penetration of av new technology or energy resource. Initially, the rate of uptake is low, with only a few early adopters. The market share may either grow gradually or swiftly (low or high value of k) approaching some ultimate saturation level. The uptake of hybrid vehicles is an example, with the ultimate market share of less than 5% in most countries.

Reaching a natural constraint in a renewable resource fits the logistic model, where exponential growth occurs until a limit is reached, causing a plateau with no further growth. As an example, consider ethanol production in the United States (RFA 2015). In 1980, ethanol production was 5.5 million barrels (Mbbl), and by 2011 had reached 442 Mbbl. The data fits an exponential growth rate of about 15% over that period, but then the production reaches a limit, mainly due to a constraint in the production of the corn feedstock. All of the surplus corn in the United States, around one-third of the crop, was converted into ethanol rather than being exported, and no further growth has occurred, even though legislation enacted in 2007 mandated production of 706 Mbbl by 2016. The legislation anticipated that the next generation of ethanol from wood or algae would be available to sustain the growth. The US ethanol situation is a good example of how past growth should not be assumed to predict future production or demand without realistic engineering analysis of the limiting factors and technology potential.

Example 4.2

Step 1 – Finding historical data for personal travel in Beijing.
  The population of Beijing 100 years ago was 2.5 million; it had been around that size for more than six centuries. The city was largely within the historical defensive wall extending 8.5 × 6.5 km, where the 2nd Ring Road is today. The activity systems of most people were within walking distance of their homes, as shown in Figure E4.1.

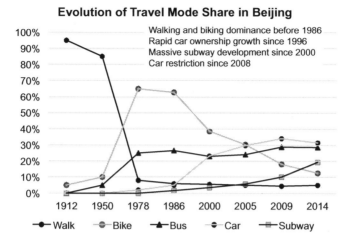

FIGURE E4.1 Personal transport travel mode share over the past century in Beijing, China.

The People's Republic of China was founded in 1949. Large manufacturing industries were set up outside the city wall, and worker's housing areas called *Danwei* (or 'work units') were designed using the Soviet high-efficiency housing model. They were purpose-built so that the entire activity system was within walking or biking distance of the assigned apartments. Families of factory workers would be assigned a residence in the *Danwei*. In the 1960s, the city wall was torn down, the 2nd Ring Road and subways were built, and the city started to expand, reaching 7.5 million inhabitants as a result of rural immigration.

The bicycle became the transport mode for more than half of all trips, and buses started to connect the new suburb areas to the city centre. After the decade of the Cultural Revolution ended in 1976, the policies of opening up to more manufacturing saw increased expansion of the city, to population of over 9 million, in the 1980s, when the 2nd Ring Road was finally completed. The additional 3rd and 4th Ring Roads expanded the city to 18.5 × 18 km.

In the 1990s, the government pursued a more liberal market approach to development, encouraging car ownership and working to modernize transport by removing bike lanes and discouraging bicycles on roads in an effort to reduce vehicle congestion. The severe congestion and air pollution issues have led to restrictions on car ownership and use since 2008. To alleviate congestion, the government is adding to the 15 rapid transit systems: 18 new systems are under construction and 20 metro lines are planned. Residential districts are now 20–30 km from the city centre, and people are largely able to move to any area where they can afford an apartment, which often means in an outlying suburb.

The current population is over 21 million. Ownership of a private car has strong status appeal, but the fastest growing market is for electric bicycles and scooters with myriad styles available. The government is now planning to increase the bicycle infrastructure and encourage bicycle and e-cycle use. The history of personal transport in Beijing shows the role of city size and organization of housing and the strong impact that government policy can have.

4.3   STEP 2: TAKE STOCK

The second step in the InTIME method is to take stock of the current energy use, policy environment, economics and environmental impacts of the activity system. This includes gathering data about population, land use, energy supply, water use and other factors that might be important in the context of the wicked problem. Taking stock also includes evaluation of risks of unsustainability. As an example, consider the wicked problem of coal mining in Australia. You can easily find data for the Australian population, land area and energy use from Australian government statistics. Australia is a signatory to the international agreement on climate change, the COP21, aimed at reducing coal use by 80% within 30 years. In 2001 the Australian government introduced a mandatory renewable energy target of about 5% of electricity generation by 2020. Australia is the largest coal producer in the world. In Australia, mining consumes more energy than all residential uses, and five times more energy than the agricultural sector (OCE Australia 2015). You can also find articles about some of the social and environmental costs of the coal mining and exporting industry, such as damage to the World Heritage Great Barrier Reef (Flannery 2014). Considering the profits flowing into mining companies, the jobs in mining and the issues of environmental costs, it is clear that the situation of coal is a wicked problem for Australia.

4.3.1   METRICS OF THE CURRENT SYSTEM

Economics is a social science concerned with the value of goods and labour, the balance between production and consumption, and the transfer of wealth between members of society. Growth in production and consumption occurs if the supply of the money, commodities and labour is in surplus and infrastructure is sufficient to support trade activities and access to markets. Economics necessarily has a short-term perspective. Economic metrics such as profits and expenditures, incomes and taxation, supply and demand, dominate the way we understand the present. However, we need to keep in mind that economics has never been an accurate way to explore the future. Economic metrics like the percentage of workers unemployed or the GDP are a distillation of the complexities of the physical and human activity systems. Think of economic measures as similar to taking your temperature. Your body works quite naturally to regulate your temperature. If your temperature drops below the equilibrium point, your body has automatic responses to try to recover, and other people can help by warming you up. If your temperature is high, it could be because you are sick and your body's immune response is fighting off an infection, or it could mean you are exercising to stay fit. Measuring your temperature is important, but it does not give the whole picture; we would need more observations. The history of growth has led to expectations that increased consumption year after year is good and necessary for prosperity. The problem is, economic growth is not the whole picture; we need more observations.

Policies, rules, regulations and laws apply governance to the system. Policies change over time and are important for setting minimum standards for efficiency and emissions. Policies can affect the economics of the system by imposing taxes or providing subsidies. Policies also support certain activities by investment of public funds in infrastructure such as bike lanes or fibre optic cable.

Assets are the buildings, roads and other infrastructure and durable goods that have already been built and are already in use. The sunk costs are the investments made on the current infrastructure and built environment. The metrics for the current assets are the age and condition of physical plant and buildings, the security of supply chains, the sources of energy and the energy intensity and efficiency. Governments have asset management programs that can provide information. Many countries have available data on building ages and performance, for example, the National Australian Built Environment Rating System (NABERS 2017). The regional assets and resources should be explored, and land use, geography, and the locations of homes and activities mapped out. Energy and activity system audits should be carried out on the system under study. The perceptions and attitudes of stakeholders are important but difficult to quantify.

Resilience is the ability to return to the previous level of activity after a disruption. Resilience mitigates risks in the short term. Resilience is high when people have options and alternatives and can easily adapt to changes. Resilience is highest when the normal activity system has low energy intensity regarding electricity and high-quality fuels. As an example, consider the resilience of household heating to a gas supply disruption. A poorly insulated home would drop in temperature quickly and would have almost no resilience, with recovery to normal activities only after the disruption is over. If the house is well insulated, has passive solar design, and has low energy demand to begin with, then the household may be able to ride out the disruption in relative comfort. If the home has a wood log burner and a wood supply, then its residents' resilience would be very high because of the availability of energy-switching options.

Adaptive capacity is the ability to use less energy to carry out the same activities without losing the essential benefits of the activity. Adaptive capacity mitigates risks in the long-term. With the residential heating example, the well-insulated passive house with ability to use wood would be able to adapt to 80% less fossil fuel use without changing their living quality. On the other hand, an 80% reduction for a poorly insulated house without passive solar or wood would have an 80% reduction in comfort quality, and thus they do not have the adaptive capacity to achieve the required fossil fuel reduction without an energy transition of the structure.

4.3.2   EVALUATE THE BIOPHYSICAL ECONOMICS OF THE SYSTEM

The term *biophysical economics* means the accounting of the flow of energy or materials through the economy rather than the money spent. The cost of a unit of energy can vary greatly over time due to any number of factors. However, the heating value of coal or the usefulness of a kWh of electricity does not change. Energy return on energy invested (EROI) helps us to understand the current and future benefit to society of different energy resources and conversion technologies. EROI will be covered in more detail in Chapter 7, but the basic concept is illustrated in Figure 4.5.

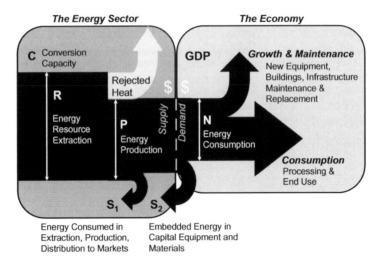

FIGURE 4.5 Definition of terms and schematic diagram of the energy system in biophysical economic terms.

The main point of evaluating EROI of a given energy supply system is the same as for financial analysis of different investments. EROI lets us compare different energy transformation platforms and make informed investment decisions based on the ratio of energy benefits to energy costs:

$$\text{EROI} = P/(S_1 + S_2) \tag{4.4}$$

where P is the primary energy production in units of MJ or MWh,

S_1 is the energy purchased or diverted from production for primary production and distribution,

S_2 is the embedded energy in the production plant and operating equipment in consistent units,

The objective of investing is to realize a return. In the energy system, the profit for the society as a whole is the net energy yield, N:

$$N = P - (S_1 + S_2) = P(1 - 1/\text{EROI}) \tag{4.5}$$

An EROI over 20 means that the energy sector is a minor consumer of energy, and the bulk of the useful energy can be used to provide services or build new buildings or products. Basic requirements for human activities and ecosystems indicate that EROI below 10 means that it is difficult to have surplus energy to spend on growth and maintenance. With an EROI of 5, the energy sector alone consumes 20% of all the available energy, making maintenance of infrastructure difficult and prosperity unlikely. EROI of 2 means that the energy sector consumes half of the energy in the economy; an EROI of 1 means that there is no economy at all as there is no energy for the economy. Producing more energy at a lower EROI results in a less prosperous economy and reduced quality of life because the energy-related resource extraction becomes one of the main activities of the economy (Lambert 2014).

Building a coal or hydropower plant, or drilling oil wells pays back the energy used within a very short timeframe compared to wind or solar. The surplus then is available to the economy and drives all other activities and growth. Historically, crops were the main source of energy as food for animals and people. Modern agriculture has an EROI over 30 for corn where the energy is in food, but EROI for corn ethanol is 0.7–1.5. This is due to the very high energy input per unit of fuel needed for distilling the water out of the alcohol. Thermal coal has EROI as high as 90 when used as heat, but EROI for coal-fired electricity generation is around 10. EROI for renewable energy conversion is highly sensitive to the lifetime of the converter components, the conversion efficiency and the utilization factor of the technology.

The key economic measurement is GDP, which is a bulk measure of all spending. The GDP increases if a large expenditure is made on solar PV or bioethanol production plants. However, the ability of those energy sector investments to support end-use activities, production and construction are vastly different than the investments in hydropower or oil production. Investment in low-EROI renewable energy production can actually cause a drag on the activity system compared to investment in conservation of high-EROI energy resources. High-EROI energy supply that is not used due to conservation by one customer is then available for other activities. Thus, the net energy available increases with conservation and declines with investment in low-EROI energy production.

The current rush to develop shale gas, oil from tar sand, or first-generation biofuels from food crops may produce short-term financial gain for speculators, but these investments create an energy drag on the economy. Low-EROI energy production schemes require a large input of high-EROI energy now, then produce a lower amount of energy over time than if the high-value resource had been used directly. In the case of residential solar PV, the intermittent nature of the power generation places operational strain on the grid system that was designed for central generation and step-down distribution and load-following rather than production-following controls. Large-scale deployment of intermittent solar PV generation requires construction of more grid infrastructure, further reducing the overall system EROI. Energy storage in batteries or conversion to hydrogen by electrolysis have negative EROI, because they do not actually produce energy. Thus, renewable energy generation with battery storage has much lower EROI than direct use of the renewable energy (Gupta 2011) (Table 4.1).

Evaluating energy efficiency investments is difficult using the same rationale for EROI as for energy supply investments. Efficiency improvements or retrofits of buildings can be made to use natural daylight, passive ventilation, natural cooling and solar heat. The return on conservation and demand management investments are generally much better than for alternative energy sources particularly if the energy production they do not use is from a high-EROI source. Energy reduction often has added benefits like emissions reduction and the improvement of resilience. But reducing demand does not generate energy, so the EROI does not apply.

TABLE 4.1

Energy Return on Energy Invested Range for Energy Production Systems from Various Resources Using Different Technologies

Energy Fuel and Conversion System	EROI Range
Surface-mined coal used for process heat	20–100
Hydropower	11–267
Coal-fired steam Rankine power plant	7–20
Geothermal steam and organic binary power plant (with co-generation)	3–39
Coal steam Rankine power plant with carbon capture and geologic sequestration	4–6
Utility-scale wind turbines and farm	7–47
Utility-scale solar photovoltaic electricity	3–6
Tar sand oil strip-mined and refined liquid gasoline and diesel fuel	3–6
Hydraulic fracturing (fracking) oil and gas fuels	<1–3

The GDP will decline if efficiency or conservation cause reduced energy consumption, unless people spend the money on consumption of other goods. It is important to understand the role of changes that result in conservation and simplification and how they extend the life of high-EROI energy conversion resources.

4.4 STEP 3: EXPLORE THE FUTURE

The third step involves exploration of different future development directions by using scenarios. Energy scenarios have been put forward by numerous organizations over the past decades. (Scenarios are also called pathways or roadmaps.) There are no established standard methods for scenario analysis like there are for engineering analysis and modelling. The basic approach can be described as storytelling based on a set of assumptions, producing an if-then narrative. Scenarios are prone to giving the results desired by the funders of the work. Non-engineers have carried out nearly all published future scenarios, including scenarios that involve engineered systems and technology.

We will first review some of the most quoted scenarios and look at their assumptions. These are typically business-as-usual (BAU) scenarios that project the growth of future energy demand. Figure 4.6 illustrates how scenarios are used to explore the different development assumptions about energy technology options. A new type of scenario called the retro analysis takes a different approach of placing the technology development objective at a point in the past to evaluate the impact. Three simple ways to carry out engineering scenarios are called *wedges* because of the pie shape they form with the BAU scenario (Socolow et al. 2004). We will add risk assessments to the results of existing scenarios by applying constraints on resources or emissions to evaluate the boundary of unsustainable development. Later in the chapter, we will use the constraints on emissions and resources to map the Forward Operating Environment (FOE).

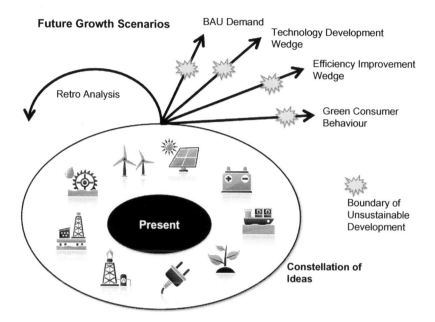

FIGURE 4.6 Future scenarios for possible developments explore the effects of different assumptions and pathways for different uptake from the constellation of ideas.

4.4.1 Review of Published Future Scenarios

It is important to note that nearly all scenarios are developed using methodologies that project future demand growth largely from past demand growth. Transition Engineering recognizes that we are currently at a historical energy inflection point, and the future context of energy production and use is not the same as the past.

4.4.1.1 Endogenous Demand Scenarios

Most governments and international organizations develop scenarios of future energy demand. The International Energy Agency (IEA) *World Energy Outlook* is the most widely quoted source of future energy demand. Shell Oil Corporation as well as British Petroleum also produce energy scenarios projecting energy demand. When the production capacity of consumer energy is much larger than the demand, then the price of energy is low, and there is potential for demand growth as consumer technologies are developed, vehicles get larger and more powerful and house size increases. However, during this century demand is transitioning from growth to decline as new conditions arise, including physical constraints on energy supply and emissions, higher energy prices, higher housing costs and saturation of improved quality of life with increased energy use in societies with high energy intensity.

4.4.1.2 Futuring Political and Technology Scenarios

The field of futuring develops scenarios for various political and technical developments. For example, the World Future Society (WFS) develops what are known as storyline scenarios using the Delphi method, which involves interviews with numerous experts and analysis of the collective expert opinion. The 2008 State of the Future 2020 Global Energy Scenarios (WFS 2008) describes four scenarios that all start with the basic premise that future energy demand will continue to grow and will always be met.

- Scenario 1. Business as Usual: current trends continue with no major change or surprises.
- Scenario 2. Environmental Backlash: environmental movements become more organized and sue for action in the courts, and some attack fossil energy industries.
- Scenario 3. High-Tech Economy: technological innovations accelerate and have an impact on energy supply and consumption with a magnitude similar to the Internet's impact in the 1990s.
- Scenario 4. Political Turmoil: the number of conflicts and wars increases, several countries collapse, migration increases, global political instability increases.

4.4.1.3 Climate Change Mitigation Energy Scenarios

Anthropogenic greenhouse gas (GHG) emissions are produced by all current industrial systems. The IEA *Bridge Scenario* gives five key policy measures that all governments should implement in order to ensure that the peak in carbon emissions occurs no later than 2020 (IEA 2015):

- Increase energy efficiency in industry, buildings and transport sectors
- Reduce use of least-efficient coal-fired power plants and ban new construction
- Increase investment in renewable energy technologies to $400 billion in 2030
- Gradual phase-out of fossil fuel subsidies to end-users by 2030
- Reducing methane emissions from oil and gas production

The concept of stabilization wedges was proposed by economics researchers as a way to illustrate how the scale-up of 15 currently available technologies or end-use changes could grow linearly from zero deployment today until each one accounted for 1 Gigatonne of solid carbon (1GtC/yr) of reduced carbon emissions in 2054 (Pacala and Socolow 2004). Each wedge was assumed to be technically feasible in 2004 and would require a particular effort to realize the 1GtC/year of mitigation by 2054.

4.4.1.4 Science for Energy Scenarios

Scientists from a wide range of fields have developed scenarios about the future of energy systems (Science for Energy Scenarios 2014, 2016). The science-based scenarios use an if-then approach. The historical behaviour of the system is used to model the system dynamics, and different if-then conditions are explored. Climate scenarios show that emission of 500 Gt CO_2 would push the climate into high risk for catastrophic change. In the timeframe to 2050, this essentially requires the curtailment of fossil fuel production. Engineering and economics researchers have applied the climate risk condition to explore different scenarios. Essentially, all of the if-then scenarios based on science indicate significant decline in demand for fuels and electricity to 2050 and beyond.

- If all technically feasible renewable energy were developed, then massive reduction in demand for transport fuels, heating fuels and electricity would be necessary.
- If all copper, aluminium, lead and steel were recovered and recycled, then demand would decline.
- If all uranium and nuclear power capability were deployed, then demand for electricity would decline.
- If the carbon market were perfectly efficient, then demand for electricity and personal vehicle transport would decline.
- If the most advanced engine technologies were deployed, then travel demand would decline significantly.
- If the full capacity for electric personal vehicles was realized, then travel demand would decline significantly.
- If all of the solar and wind capacity possible were developed, then the electric grid would require double the capacity, storage would be required, cost would rise and demand would decline.
- If all of the renewables possible were developed, then the lower EROI would result in reduction of net energy production and demand would decline.

4.4.1.5   Sustainable Energy Scenarios

Sustainability assessment and sustainability indicator methodologies have been developed over the past decades. A sustainability indicator is a way to communicate the relative merits of a development pathway considering environment, economic and social factors (Singh 2012). Numerous scenarios have been constructed with different methodologies, with the underlying thrust of achieving energy sustainability. It is hard to find any that recognize the fact that the energy transition requires lower energy demand overall. For example, the World Energy Council (WEC) scenarios aimed at reducing carbon emissions still show growth in primary energy production on all continents (WEC 2013). The Bruntland Commission definition of sustainable development has never been disputed or even debated (WCED 1987). However, it should be obvious that irreversible impacts and consumption of finite, non-recyclable resources pose a risk to future generations' ability to meet their needs.

4.4.2   ENERGY TRANSITION SCENARIOS

The Transition Engineering approach is different from conventional scenarios that predict future demand. We flip the perspective to look at fossil fuel production and use rather than focus on the resulting carbon emissions. The biggest contributor of CO_2 emissions is from burning fossil fuels for power generation, manufacturing, cement manufacture, materials processing and vehicles. Engineers are responsible for extracting and delivering energy resources to the market. All end-use facilities and appliances are designed and manufactured; therefore, all emissions from fossil fuels were designed into the system. The scientific community, the public and the media are now looking to policy-makers to set targets that will mitigate the risks. The InTIME methodology explores four scenarios with the aim of quantifying the unsustainability risks to the essential activity systems for different trends into the future.

- BAU: The current trend of the end-use activity and the energy supply continues to grow at historic rates.
- Technology Wedges: The end-use activity continues on the current trend, but new end-use technologies and energy resource conversion technologies develop and grow market share, resulting in fossil fuel demand reduction.
- Efficiency Wedges: The end-use activity continues to grow at historical rates, but new efficiency improvements grow in market share, resulting in lower fossil fuel demand compared to the BAU path.
- No Growth Virtuous Consumption: The current level of energy activity and energy demand continues at current levels without growth. This scenario depends on virtuous consumer behaviour, where people choose lower carbon products, behaviours and lifestyles.
- Forward Operating Environment (FOE): For the end-use activity, FOE is an envelope of probability of the physical availability of the resource or pollution sink availability into the future. Figure 1.4 in Chapter 1 can be used to develop the FOE for conventional oil for a nominated probability. The Representative Carbon Pathway (RCP) for global warming forcing of 2.6 W/m^2 (RCP2.6 Scenario) is the FOE for having a better than 50% chance of global temperature increase remaining below 2°C (IIASA 2009).

Figure 4.7 illustrates the forward risk environment for different scenarios and the transition pathway. The historical data is indicative of oil production, but the same type of graph could be used for emissions, water use, tons of plastic in the ocean, and so on. The scenarios are defined for a specific geographic location and a particular end-use activity. One or two important energy parameters are chosen to characterize the demand along a future scenario trajectory. The objective of the scenarios is to estimate the timeframe for the unsustainability of the activity to cause a change.

Semi-quantitative risk assessment (SQRA) is well known in civil and natural hazards engineering for making estimates of the probable loss of life or property damage due to flood events, like 100-year floods, or earthquakes. The likelihood of events is laid out as rows in a table, and the severity of impact is laid out as the columns. The highest risk is for likely events with major consequences. The increased risk from global warming is due to the increased severity of weather-related hazards like hurricanes, floods, droughts and fires, and the increased probability of occurrence of severe storms. As a power plant ages past its design and material life, the

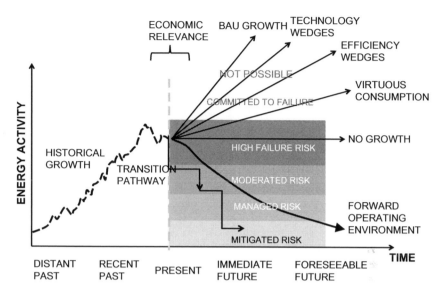

FIGURE 4.7 Future scenarios move the energy, environment and economic activity system into different risk levels. The climate change mitigation path for nearly all activities other than environmental restoration require management for consumption decline. (Reprinted by permission from Springer Nature: *Transition Engineering* by Krumdieck 2017.)

risk of a fault or accident increases. In the case of nuclear power plants, the potential scale of a catastrophic failure also increases with the age of the plant. The risk of failure increases to unacceptable levels as a resource is depleted if the end-use activity that uses the resource does not adapt to reduced supply. This is a key concept in Transition Engineering: a depleting resource is sufficient if the end-use demand reduces faster than the resource.

4.4.3 BAU GROWTH FUTURE SCENARIOS

Even if you are sceptical that past growth will continue into the future, it is hard to reconcile our understanding of unsustainable growth with the fact that, indeed, tomorrow will largely be like today and business as usual will continue. Use Equation (4.1) and spreadsheet analysis to estimate the historical growth rate for the energy activity and to project the future BAU scenario. Use the other growth models to explore linear growth and logistic growth scenarios. What is the historical doubling time? What are the physical limits on the system? When will they be reached if BAU growth continues? The main finding of the BAU scenario is to understand the timeframe within which BAU will change because it cannot continue.

In 1973, New Zealand had 4,000 MW installed electric-generating capacity that was 100% renewable, 2,000 MW of glacial lake-fed hydro on the South Island, 1,900 MW of run-of-the-river hydro on the North Island, and 100 MW of geothermal energy from the volcanic zone in the North Island. Consumption was only about 20,000 GWh, or: 20,000 GWh/(4,000 MW × 8,760 hr) = 57% utilization

In 1982 consumption had grown to 25,000 GWh and was growing at a rate of nearly 5% per year. At this continued growth rate, the consumption would have exceeded the renewable generation capacity in just seven years. If the growth in consumption had continued at that rate, the country would have needed to find a whole second South Island by 1998. The growth rate did slow to 2% after 1986, and has been around 0% since 2006. And the country did not find another South Island; it built a 1,000 MW coal power plant and almost 2,000 MW of natural gas power plants.

Limits to growth are difficult to comprehend during the growth phase of the depletion curve. However, the ecological and material limits to growth are obvious in retrospect, after the boom and bust. The island of Singapore, according to World Bank statistics, had steadily growing energy consumption per capita until 1990, then it grew at an average rate of 18% per year until 1994 and has remained in an undulating plateau since then. During the period of rapid growth, future projections showed continued demand growth beyond 2020. The limit to the demand growth on the island of Singapore turned out not to be the limits of its citizens' ability to purchase fuel; it was the limits to how many buildings and uses of energy could be added. The government also instituted strong efficiency and conservation measures. Thus, the 1990s BAU did not continue, even though at the time it was expected to.

4.4.4 MITIGATION WEDGES

In the historical study of supply and demand, we observe that demand grew because supply grew. Our logical conclusion is that if the supply declines, then the demand will decline to follow it. As evidence, consider the Organization of Petroleum Exporting Countries (OPEC) oil embargo of 1973–1974. The supply of oil in the United States declined, and so did demand, even though Americans did not want to consume less oil.

Most scenarios for sustainable future development involve the proposed uptake of technologies that change the unsustainable aspects of current activity systems without actually changing the systems themselves. The International Panel on Climate Change (IPCC) and the IEA both rely heavily on the assumed viability of carbon capture and sequestration (CCS) for future scenarios in order not to exceed 2°C overall global warming. The IEA estimates that CCS will need to meet one-sixth of all necessary carbon reductions by 2050 (IEA 2015). James Hansen, leading climate scientist from NASA, recommends that nuclear energy would be the way to reduce carbon emissions enough to stay below the high-risk warming (Hansen 2009). Biologist Tim Flannery explains the effect on species and ecosystems already being observed and recorded around the world, and recommends that renewable energy, and in particular solar energy, is the answer (Flannery 2005). Here is a spoiler alert: our analysis of alternative energy technologies will show that the energy transition is from high fossil fuel use to low fossil fuel use, *and* thus low energy demand.

Is a proposed technology really a viable part of the energy transition? The InTIME methodology uses the technology development vector analysis (described in more detail in Chapter 5) as a tool both to assess the probability of a green energy idea being realized and to communicate the results of the analysis to decision-makers and the public. A technology idea should not be of interest to anyone if it is not scientifically reasonable, technically feasible and economically viable. However, in the current world of social media, unproven energy technology ideas are found in abundance circulating on the Internet. One example is the 'solar road', which raised significant development capital on claims that solar PV panels embedded in road surfaces could address the need for green electricity.

In 2004 the Carbon Mitigation Institute at Princeton University published the idea of carbon stabilization wedges (Pacala and Socolow 2004). The stabilization wedges were technologies that could be developed over time and envisioned to reduce carbon emissions by about 7% in 2054 from a BAU projection that would otherwise reach 14 GtC/yr. The 15 wedges are given in Table 4.2 along with an estimation of the probability that the 2054 mitigation measure could be achieved. IPCC and IEA have not added any new strategies to the potential wedges proposed by Pacala and Socolow, no new technology breakthroughs have occurred in the years since the carbon stabilization wedges were proposed, and the viability has not improved for the technologies seen as carbon mitigation wedges.

Lifecycle assessment (LCA) is a well-known analysis used to compare the energy or material intensity of different products. The process involves tracking the inputs and emissions from each step of the manufacture, distribution, use and disposal of the product. LCA analysis can be quite difficult in the current context of global supply chains and market distribution. Many LCA studies have been carried out, for example, to compare the GHG footprint for electric cars to compact diesel cars. LCA is a reasonable approach for comparing products like-for-like. However, it is difficult to use LCA to evaluate the sustainability of the whole system, for example, personal vehicle-based transportation in low-density cities. LCA and EROI should be used to help evaluate the probability that a new technology or end-use change will be developed along the future scenario trajectory.

Most people believe that technological progress is needed to satisfy our growing hunger for energy. Most people hope that technology solutions are being developed by researchers at universities and by entrepreneurs who have a vision of a sustainable future. Many people get excited about new clean energy innovations. We all agree that it is worth trying ideas that are not currently economical because economies of scale will bring down the costs in the future, and even trying something new creates by-products.

A few key words in the preceding sentences are important to consider when you are gathering information and data about a green energy idea: *belief, need, satisfy, growing hunger, hope, solutions, vision, excited, clean energy.* These types of words are evocative. They trigger emotional responses that can be used to mask issues with the proposed technology idea. The technology development vector analysis is

TABLE 4.2

Carbon Stabilization Wedges Proposed by Pacala and Socolow in 2004 and a Probability Estimate That Each Could Be Scaled Up to Produce 1GtC/yr Emissions Reduction by 2054

Mitigation Measure	Scale-Up Effort by 2054	Possibility of Full Scale-Up by 2054[a]	
		Feasible Technology Development	Viable Adoption and Market
Efficient vehicles	Replace all cars (2 billion in total 2004–2054) with 60 mpg fuel economy	100%	70% (30% for larger vehicles)
Reduced vehicle use	Cars have 30 mpg average miles per year reduced 50% (from 10,000 mi/yr)	100%	100% if urban form changes
Efficient buildings	25% reduction carbon energy use	100%	100%
Efficient baseload coal plants	Doubling of coal power generation, but all plants generate at 60% efficiency (from 32% today)	20% 40%–45% efficiency is feasible for coal	20% Coal plants increasingly uneconomical
Gas substituted for coal in baseload generation	Four times current production with gas (1400 GW)	20% Gas supplies in steep decline by 2054	60% Where gas is available at low cost
CO_2 capture and storage (CCS) at power plant	800 GW of coal and 1,600 GW natural gas built with 100% CCS	0%	0% Cost prohibitive
CO_2 capture at H_2 plant	250 MtH$_2$/yr of hydrogen production from coal plant with CCS	0%	0% No market for H_2
CO_2 geological storage	Create 3,500 'Sleipners' storage in depleted gas reservoirs (currently 1)	0.03%	0% CO_2 injection for Enhanced Oil Recovery (EOR)
Nuclear power substitution for coal	Add 700 GW nuclear power generation (twice current capacity)	0% No technical options for waste storage	0% Decommissioning without replacement
Wind generated power substitution for coal power	Add 2,000 GW capacity (47.6 GW in 2004, 370 GW in 2014)	80% Rare earth minerals Wind is not baseload	80% Requires demand side response
Solar PV substitution for coal power	Add 2,000 GW capacity (3 GW in 2004, 229 GW in 2015)	80% Glass, silver materials; solar is not baseload	80% Requires demand integration
Wind generated H_2 fuel cell cars substitute for gasoline	Add 4,000 GW capacity, H_2 storage infrastructure and Fuel cell vehicles	0%	0%

(Continued)

TABLE 4.2 (*Continued*)

Carbon Stabilization Wedges Proposed by Pacala and Socolow in 2004 and a Probability Estimate That Each Could Be Scaled Up to Produce 1GtC/yr Emissions Reduction by 2054

Mitigation Measure	Scale-Up Effort by 2054	Possibility of Full Scale-Up by 2054[a]	
		Feasible Technology Development	Viable Adoption and Market
Biomass fuel substitution for oil	Increase current bioethanol by 100 times, using one-sixth of world cropland	0% Maximum in 2015	0% Food more valuable
Forest restoration	Decrease tropical deforestation to zero, reforestation on 300 Mha	100%	100%
Conservation Tillage	All croplands (increased from 10% of cropland in 2004)	100%	100%
Low CH_4 Diet[a]	60% reduction in cow numbers, 30% reduction in pigs	100%	100%
Industrial activities[a]	Reduce construction and manufacturing by 50%	100%	100%
Leaks and pollution[a]	Eliminate all CH_4 leaks in gas, oil and coal production; capture and burn all CH_4 from landfills and sewer	50% Widespread	30%

Source: Excerpted from the Carbon Mitigation Institute at http://cmi.princeton.edu/wedges/.
[a] Added by Krumdieck (The Shift Project 2010).

a way to separate the emotion from the engineering science and data. The analysis recognizes that there are different types of hurdles to be overcome by research and development for any idea to become a viable product meeting a need in society. If there is even one hurdle that cannot be overcome due to unachievable fundamental science, energy balance or material properties, impractical manufacturing or application issue, then the technology will *never* be available in the market.

Energy technologies that actually require more energy than they produce cannot provide a viable energy source for an industrial society. Examples are hydrogen, most ethanol biofuels, conversion of natural gas to methanol, capturing CO_2 and converting it to liquid fuel and beaming solar energy to Earth from space arrays. Batteries are often referred to as clean energy technology, but they are not an energy supply technology, and they impose an energy penalty of 5%–20% on the electricity system.

As a final example, we will examine why there will never be a transition from petroleum fuels to hydrogen for private vehicles. Hydrogen production, storage and distribution has insurmountable hurdles at every stage of the development vector. In addition, the hydrogen transition would require an even larger investment in infrastructure than for petroleum, but for a fuel with a net negative EROI rather than for an

EROI greater than 20 on which the petroleum system was built. The use of hydrogen as a transport fuel would accomplish the same end-use services as using petroleum; that is, there would be no marketable benefits associated with the enormous costs. Finally, at least 30% of private vehicle trips are optional. They could be carried out with active modes or public transport. 50% of vehicles are larger and use more fuel than needed to accomplish the activity. As conventional fuel and resources for building private vehicles and road infrastructure stops growing and starts to decline, car travel will become more expensive. It is much more likely that people will do sensible things that reduce costs rather than spending more to do the same unnecessary things.

4.4.5   No-Growth Scenario – Virtuous Consumption

The no-growth scenario is not likely to be an option considered by economists or policy-makers. However, since the 1990s, there are many examples of energy activity systems that have not experienced growth. For example, Hong Kong's per capita energy use has not grown since 1998, despite the past two decades being a prosperous time for the city. Most European countries have had no-growth or declining total energy demand for the past three decades (BP Statistics Review 2017). No growth can be prosperous as long as the society is not in debt. Debt requires increased income in order to pay down the debt and provide the income for continued operations. Throughout the last century, especially in the United States, Canada, Australia and other low-population-density areas, expenditure on new ventures and new infrastructure had the potential for great returns in growth in economic activities. Thus, a city could issue municipal bonds to pay for a new library because there would be new houses and new businesses into the future, which could pay off the bonds as well as keep the police and water workers paid. The field of economics emerged during this unique time of unprecedented growth in population and in building, extraction and production. There are many leading economists who are now defining the problem of understanding no growth and negative growth and working out how to be prosperous as a society under these circumstances (Jackson 2009).

4.4.6   Forward Operating Environment (FOE)

The general attitude toward climate change appears to be that reducing fossil fuel use is primarily a consumer option or would occur as a consequence of increased renewable energy use. This attitude leads to a passive approach to fossil fuel and GHG-producing activities. In contrast, the InTIME methodology sets 80% GHG reduction by 2050 as a requirement. The unfolding constraint envelope on consumption or emissions is called the FOE.

Consumption growth is not sustainable under any scenario. Growth is always limited, temporary and inevitably followed by decline. All resources are finite and limited. All growth requires resources. Thus, exponential growth in energy and resource use always has a peak and subsequent decline associated with one or more depletion effects. The crude oil that has been used in every aspect of the economy over the past century is clearly a finite resource. This *conventional* crude oil has the highest EROI and is the easiest to extract and refine. Since the oil price increased in 2008 and 2009, lower-quality hydrocarbon sources have been rapidly brought into production. These 'high-hanging fruit' include tar sands in Canada, tight shale oil in the United States and several other countries, and deep sea and sub-salt offshore oil.

Figure 4.8a shows two different production curves for the same 100-unit reservoir, produced using Equation (2.6). The difference between the curves is the production growth rate. Experts may not agree on the total amount of a resource, and people rarely think about the lifetime depletion of a resource, so using the Hubbert model to make a graph of the resource production lifetime can help with scenario planning. Figure 4.8b shows the total production from four oil reservoirs developed over time. The first reservoir is 100 units, the second reservoir is 50 units, the third is 20 units and the fourth is 10 units. The peak in total production is also the peak of the major reservoir. Development of the smaller reservoirs maintains the production at an undulating plateau for 20 years, followed by a steep decline despite the development of two further resources.

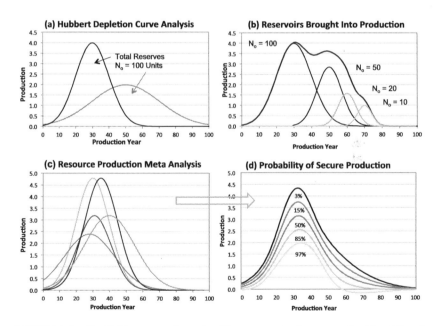

FIGURE 4.8 The Hubbert curve is used to illustrate production curves for (a) a 100-unit reserve with peak at 30 years or 50 years, (b) four reserves of diminishing size brought into production when the previous reserve reaches peak and the total production curve, (c) production curves for the same resource produced by five experts based on different assumptions, (d) the probability of production in a given year developed from statistical analysis of the five expert analyses together.

People who are specialists in the geology, extraction and markets for different resources can use different data and assumptions to produce an estimated production curve, as illustrated in Figure 4.8c. If we have a number of experts in the field producing different estimates of the production curve, then we use the meta-analysis to include all of the expert combined wisdom. Figure 4.8d is an example of a meta-analysis of experts with probability of production in a given year derived from the mean and standard deviation of the expert production curves. The lowest curve is the production that 97% of experts agree will be available. The top curve has the agreement of only 3% of experts. These simple models can be used to develop a probability estimate for the availability of an important resource into the future and thus define the FOE and the probability of future growth scenarios. This type of analysis can be understood and used by decision-makers because they can decide their risk position and then use the associated future supply estimate for long-term planning. This is especially important for planning large and expensive infrastructure to meet current needs but which may not be needed in the future. For example, consider the project of widening a freeway because of high congestion. A major highway construction requires maintenance, but it will have a physical lifetime of 50–100 years or more. If the demand for driving on the highway is likely to decline to half current levels within 20 years, then the expenditure on the expansion of the highway is not only unnecessary, it could pose a risk to the community in regards to the debt incurred and the ongoing maintenance costs. Refineries, pipelines, and factories to produce vehicles require similar consideration, as does building of outlying suburbs where residents would have no options other than driving private vehicles great distances.

A rush into non-conventional fossil energy resources has a high risk of creating a major environmental crisis. Tar sands operations are causing widespread pollution and increased cancer in local residents (Edwards 2014). Hydraulic fracturing of tight oil was exempted from the US Safe Drinking Water Act in 2005 at the behest of then vice president Dick Cheney, and has proceeded without scientific monitoring, regulation or oversight (The Editors 2011). Deep-sea oil production has been responsible for the worst oil spill disaster in history, when the Deepwater Horizon exploration well in the Gulf of Mexico blew out in 2010, killing 17 people and causing an oil-spill-affected area of 182,000 km^2, including most of the Gulf Coast of the United States (BP 2015). The tar sands, shale oil, deep sea and Arctic resources are not new resources. Rather, they are the last resources. On one hand, companies that extract crude oil will make a profit if the market price is higher than their costs to produce. On the other hand, optimism about costly extraction of smaller quantities of lower-grade hydrocarbons from more difficult locations is certainly misplaced.

4.4.7 Armageddon Scenarios Not Allowed

Climate change can elicit a negative emotional response as our belief systems and experiences are put under pressure. Peak oil is a real problem that will require large-scale adaptive change and could strand transport infrastructure assets, but the probability of running out of oil and experiencing a collapse in food supplies is reasonably low. Engineers need to recognize that they will likely have the same emotional response as other people and must manage the psychological integration process of their team. Find reliable sources of scientific information and deal with the actual data in order to accept that the problem exists and understand the nature of it. Green alternatives are fine if they actually accomplish the mission and address the problem directly.

Climate change is not a future problem. The current atmospheric loading and thermal forcing of the climate is already increasing the probability and severity of weather extremes and causing numerous crises in ecosystems and farming systems around the world. Increased CO_2 concentration in the atmosphere is causing acidification of the oceans. The only acceptable FOE is the one where the 80% emissions reduction was achieved before 2050.

4.5   STEP 4: TIME TRAVEL

The fourth step is a brainstorm process of generating path-breaking concepts about the world after a successful energy transition 100 years in the future. The general result of the future scenarios and FOE project is a new perspective, one that is a break with the historical development trend. The path-breaking concepts are normally focused on the essential requirements, and are not limited by any other current assumptions or historical context. Path-breaking concept generation requires highly creative and innovative thinking. It is not about modification of the existing system, but it does have to comply with known science, resource availability and engineering feasibility. The greatest challenges in the path-breaking brainstorming session are not to think about the current situation or the past. This is why we usually work in a brainstorming group and set out rules and correction procedures if a participant violates the rule of slipping backward.

The first step is to identify the assets in the existing systems that are highly likely to be present in 100 years. The team then accepts that the path-breaking future exists and mentally set out to discover it. The discovery is guided only by the system requirements, physical reality, and the art of engineering and unconstrained creativity.

The 100-year concept project is undoubtedly the hardest project in the framework. There is little experience with long-range design concept generation for complex engineered systems. We have a lot of people who imagine the future in the science fiction genre, but where can you find reasonable and realistic explorations of the engineering of the next century? Authors and movie-makers often use the distant future as a setting for exploring how dangerous trends might evolve. Sometimes the engineering isn't very believable, but sometimes, the technology and the way it can all go wrong seem too believable for comfort. In the story *1984*, George Orwell creates a future that appears to be utopia which is actually a dystopia where media, communication and surveillance technology allow Big Brother to control thought as well as behaviour.

In 2003 I asked an audience at a public lecture to fill out a simple survey with two questions: (1) Think of a book or movie that you think shows the probable future 100 years from now, and (2) Is it realistic and possible? The results were a bit disturbing.

1. 80% of people said *Mad Max* was the probable future. *Mad Max* is set in a post–nuclear war Australia where the oil supply is gone and civilization has broken down, although it was not directly bombed. Twenty percent of the people said *Star Trek,* which represents a more distant future where, even though there was a terrible world war, the nations of the world and indeed the different star systems have joined together in peaceful endeavour, and they have unlimited energy from dilithium crystals.
2. Unfortunately, 90% of the people who said *Mad Max* answered yes, it was possible. Only 10% of the people who said *Star Trek* answered yes, it was possible.

While these results show a good understanding of the fictional portrayal of infinite energy, it also says something disturbing about the vision of the future without oil.

Even though this was just one sample of public vision of the future, the results highlight the problem we have with realistic long-term planning. Engineering explorations of the 100-year concepts will be of great interest to the general public and could have an important influence on long-term thinking. But the possible future essential activity systems with realistic energy and built environments will depend entirely on the concept engineering.

The 100-year path-breaking project for a wicked problem is best explored in a group brainstorming workshop. Define the essential activity in the location under study, for example, the food in Naples, or personal transport in Beijing. It is important to understand that societies have always figured out how to get the essential activities done, whatever environment or resource base they happen to have (refer to Figure 4.2). For example, if the system is housing, reflect and discuss that people will always build homes to provide security. Thus, we know that the 100-year future concept has residences: places to spend the vulnerable hours of sleeping, and to care for the vulnerable people in their family group. However, the functional design of the houses should be left wide open and not constrained by the current designs. In 100 years, a substantial number of current homes could be redeveloped, so stay focused on the essential activities like hygiene, preparing meals, storing personal goods and providing shelter. The key requirement for any human society is the access to all of the necessities like food and water, and access to activities. A feedback control model of anthropogenic system dynamics will be presented in Chapter 5, which may also be helpful in developing the 100-year path-breaking concept.

The essential activity is set in the context of other essential activities, so map out the connections using the bubble map from Figure 4.2. Nominate and evaluate the elements of the technology, infrastructure and built environment that will be in the path-breaking concept. Think about the essential activity and observe the functional patterns in daily, weekly and annual timeframes. For example, if the essential activity is food in Naples, Italy, the daily calorific requirement and nutrients are known, the necessary elements of food that support the culture can be carried through the path-breaking concept. For example, olive oil, wheat, tomatoes, basil and cheese are necessary in order to cook many Italian meals. But the food infrastructure in 2120 likely will not include supermarkets with imported tomatoes delivered by diesel truck. In your path-breaking concept, observe the daily, weekly and annual patterns of life. For example, in Spain people will surely have an after-lunch siesta, and in France people will have a baguette for breakfast.

All civilizations, in all societies, in all times and places have had an economy. All civilizations in the future will have an economy. One thing that does not carry forward to the path-breaking concept is the current currency. The brainstorming group needs to use concepts of value rather than cost. For example, one day of farm labour or teaching has a real value. Food and necessary materials also have value. The path-breaking energy production systems, extraction of natural resources and production of wastes must be in balance with the rates of environmental services and within the constraints of carrying capacity. In addition, the EROI of the energy supply sector must be greater than 10.

The end of the oil age is the constraint on the FOE shaping the essential activity systems of 100 years in the future. Every system that currently relies on oil for movement of goods, workers and customers will not use oil for transport in the path-break concept. Some things will be the same in 100 years. The locations of cities, rivers and mountains, for example, can be assumed to be the same in 100 years. Major universities, government buildings, historic churches and museums can also be assumed to still be in use. Other things, like the source of mineral resources, will be profoundly different, as the recycled resources will have become more feasible than developing new primary sources. In most countries outside the African continent, the population growth rate will have slowed to near zero or will be declining (UN 2012). For simplicity we normally use the current population in the 100-year concept.

The most important aspect is to put away your own experience of the current time. Also recognize which of the future scenarios from Step 3 have low probability, and make sure to break away from those ideas for the 100-year brainstorming session. Once you set the primary suppositions for the 100-year concept, make the essential activities the focus of the concept generation. The following primary suppositions are used to start the creative process 100 years from now we know that:

- Oil is not available in quantities or at prices that support private vehicle use or transport of low-value goods.
- Coastal settlements will have retreated,
- Infrastructure and buildings durable enough to last for 100 years will still be used.
- Metals and minerals will be available primarily from recovered and recycled sources.
- The population will be no more than today, and will be declining.
- Prosperity will not depend on growth in consumption.
- Plastic will not be disposable.

It is best to use an interdisciplinary creative team together with the engineers, and have them focus on 'discovering' the ways that the essential needs of the population in the project area are prospering under the specific conditions of the FOE. Most activity systems currently have numerous wicked problems, and the challenge is to accept the premise that these are no longer problems in 100 years. Something changes now and the wicked problem is no longer the issue in 2120.

4.6 STEP 5: BACKCAST AND TRIGGER

4.6.1 WHAT IS DIFFERENT IN 100 YEARS?

The fifth step is a backcasting of the innovative path-breaking concept systems to see how much they differ from the current systems (Figure 4.9). Backcasting in product development is normally focused on assessing existing capabilities, supply chains and manufacturing plants for retooling to produce the new product. Backcasting can also identify barriers and strategies for overcoming them. For energy systems, the backcasting exercise looks at the adaptive capacity of the end-use sectors and at the infrastructure and behavioural aspects that are fundamentally different in the path-breaking concept. It then looks at how they could be modified from the present to develop in the direction of the path-breaking concept.

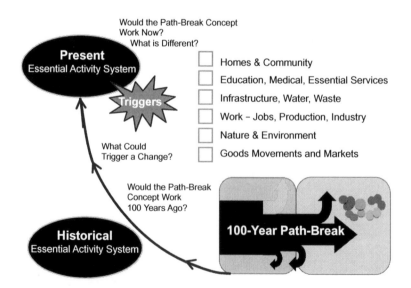

FIGURE 4.9 One-hundred-year path-breaking concept and backcasting observations lead to disruptive ideas of trigger actions.

In the backcasting project, it becomes clear that some of the currently economically successful industries and development patterns are in fact sunset industries. For example, when looking at the shape of the future city with 80% less fossil fuel and backcasting against the urban form, land use and transportation networks of the city 100 years ago, we first observe that the shape of the city from 100 years ago is still present. We discover that properties in the outer suburbs are at high risk of losing their value within the payback period of the mortgages held on them. Loss of value is a risk to residential property mortgagees since the value of a residential property is set by the willingness of a buyer to pay. The 100-year path-breaking concept has housing without mortgage payments greater than 30% of incomes, and without appreciation growth of property values. Observe the difference in quality of life when housing is accessible for all, and compare to the current situation.

Anticipate investments and developments that are *not* going to happen due to the transition away from personal automobile use. The case of urban sprawl developments in the United States is a particular example. Local governments gave the consent for new subdivisions, but the oil price spike and economic recession of 2008 has left these projects abandoned, and they require maintenance or remediation even as they bring down the value of the nearby suburbs (Holway et al. 2014). Using what we learn about the city 100 years from now, we can clearly see that continued outward expansion of housing is not sustainable and not even viable. The lessons we learn from the examination of the 100-year concept can help to avoid bad investments now. In the case of urban activity systems, we can also identify the triggers for transition, which could include oil price shocks, oil shortages and shifting preferences for living without long commutes via car (see example 4.3).

4.6.2 TRIGGERS FOR CHANGE

In engineering change management, triggers are identified early in the planning work, and there are different kinds of triggers at different stages of a project. At the start of a project, the trigger can be market research, the decision of the governing board or CEO, or the need to replace old facilities. For example, the designs, and procurement of subcontractors are planned, but construction cannot begin until the financing is approved and/or local government consent is secured. The triggers are often outside the control of the engineering team, but they are part of the project design, and the workflow is managed around the trigger events.

Energy transition triggers can be conditions or events that result in a different perspective among the general population. As an example, consider the marked effect that the movie, *An Inconvenient Truth*, had in shifting the perspective about climate change (Brulle et al. 2012). The scenario investigations, defining the FOE, and exploring the 100-year path-breaking future, give us a new perspective on the future. In the future, as the BAU progress becomes impeded and the pressure for change increases, there will be events that cause the social perspective to shift.

Energy trigger events can be caused by a new tax or penalty, such as a fuel tax or emissions charge. The 1970s energy crisis caused a shift away from diesel electricity generation and the introduction of compact, fuel-efficient American car models. Energy price increases have historically caused shifts in consumption behaviour and new markets for products with improved energy efficiency. Energy system disasters can also be triggers. It may be that the Fukushima nuclear power plant disaster will become a trigger point for a shift away from nuclear power.

The most significant triggers for the energy transition have the effect of reducing fossil fuel production. The most direct triggers are new laws that limit or prohibit exploration, extraction and transportation of fossil fuels. For example, the lack of a pipeline from the Alberta tar sands to the West Coast has limited the expansion of tar sand mining. There have previously been bans on oil drilling in environmentally sensitive areas, such as the California coast and the Arctic National Wildlife Refuge in Alaska, and a ban on deep-sea oil exploration was enacted by New Zealand in 2018. *Divestment* is a term used in business to describe when a company is changing direction and winding down part of its operations or one of its products. The global movement to divest investments in fossil fuels was started in 2011 by US university students urging their universities to shift investments from fossil fuels to renewable energy. The movement is now a global organization (gofossilfree.org) and lists 800 participating organizations and divestment worth over $6 billion. Triggers can also be new technologies or devices that enable new ways of doing business or gathering information, particularly if the trigger involves behaviour. The rental bike systems and scooters now being deployed in many cities may become triggers for wider acceptance of the bicycle as a form of transport, as could a movie in which a celebrity is portrayed bicycling.

4.7   STEP 6: DOWN-SHIFT PROJECT

The sixth step is the innovation and development of a shift project. The objective is to develop a shift project that could be presented to investors as a business proposition. A project brief should be prepared to set out the case for doing the project, the concept design, the project plan, needed the resources and risks. The introduction of the brief gives a succinct presentation of the current system, the definition of the problem, the unsustainability of BAU, and the probabilistic assessment of the scenarios that are currently being considered. Then describe the idea and how it is new, and explain how the shift would represent a transition from BAU. Also present a competitive position assessment (see Section 4.7.1) for the organization with the shift and without.

Shift projects are the *way* that the system changes to use at least 80% less fossil fuel. Shift projects are necessarily innovative because people do not yet know how to achieve the COP21 requirements. Nobody can explain how innovation happens, but the first steps are defining the problem; establishing that there are no known solutions; and creating, playing and thinking outside the box. A brainstorming workshop with teams is useful for creating a safe and free environment for trying out ideas that might seem crazy. The innovation brainstorming is sometimes called innovation games, and the atmosphere of generating ideas without judgment or evaluation is important. The InTIME team members should study the range of brainstorming techniques, try out different approaches and use what works for them.

Existing systems have great inertia to continue with BAU because of prior investments, beliefs and habits. But systems can also change very quickly through change management or failure. The shift project is the engineered change project that is designed to initiate transition. The shift project also has a learning cycle, and it is likely that the design and development of the shift project will require research into new modelling and particularly new economic models. Execution of a shift project would involve communication and gaining cooperation from stakeholders. The shift project also represents the beginning of changes in business relationships, the built environment and technology. A challenge for shift projects is not to make the project *about* a crisis. Rather, the benefits of the shift project should be obvious to all. The results of the shift project would work for people 100 years from now, and for people today.

A set of shift projects were proposed for the city of Dunedin personal transport activities (Krumdieck 2010). These projects included rezoning areas of the central business district for mixed use, apartment renovations of old historical buildings, and providing a walkable urban community for retired people. A plan for designing and implementing kid-safe cycle ways in the neighbourhoods around the primary schools improved safety and reduced car trips to the schools by 50% in five years. Another shift project set out the conceptual design for a network of electric trams that could provide 50%–70% of the access between homes and activities in the city.

4.7.1 COMPETITIVE POSITION

Social responsibility for mitigating the risks of climate change should be sufficient reasons for transition, but in the current economic context, the critical leverage for the shift project is competitive advantage. Decision-makers normally compare the net present value, payback period and rate of return as the main criteria for investment alternatives. Unfortunately, measures to 'save energy' do not compare positively with 'doing nothing'. The investment in saving energy must have a payback within a few years, while doing nothing does not have to pay back at all. This strange situation arises because the mathematics used to evaluate the present value of future benefits assumes a growing economy. This results in discounting of the future. Part of the problem with this conventional thinking is that it discounts what happens after the payback period, and it doesn't consider the decisions of competitors.

Competitive position is a new kind of transition economics that moves the perspective into the future and uses the value of assets and costs at those future times to compare between competing options. Transition economics will be described in detail in Chapter 7. The precaution here is that the evaluation of the shift project economics is not a simple matter of calculating the payback period.

4.8   STEP 7: SYSTEM TRANSITION

The seventh step in the InTIME methodology is to consider the future transition pathway that follows from the shift project. This is actually part of the shift project development process and the transition evaluation feeds back into the design. Figure 4.10 guides the discussion of how the successful shift project would disrupt the BAU and germinate new opportunities in other related systems.

The shift project results in a down-shift in energy. The shift project will create momentum change, and new ideas and new business. Carrying out shift projects may require years, depending on the scale of the changes. Shift projects may be planned for initiation when the existing system becomes uneconomical or reaches the end of life. However, the competitive position analysis may show the advantages of getting started on the transition early. The novel aspect of transition management is realizing benefits and profits in the context of curtailing the growth of finite resource consumption. New zero net energy buildings or new renewable power generation plants are considered transition projects when the overall project includes retiring of below-standard buildings or coal-fired power plants. Logically, energy systems are presently at an inflection point. Transition management is the process of changing existing systems to mitigate the range of risks that past growth has caused.

Right now, any development project that transitions from fossil fuel will provide the community or organization with the freedom to operate into the future. If the design would work 100 years from now, and if the product would last and retain

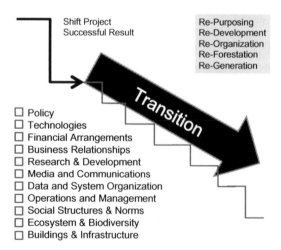

FIGURE 4.10   Evaluation of changes in related systems due to the shift project that represents transition to very low energy and carbon.

its real value for 100 years, then the real value of the project justifies the economic investment in the change. Every transition will change the story of the future. In this chapter, we have seen how important the story is. What we think is possible in the future may even be the most important factor in realizing our potential and utilizing our ability to change course and transition to the world beyond fossil fuels fast enough to avoid dangerous global warming.

Example 4.3 Shift Project and Transition Management for Oil-Based Transport

The wicked problem is personal transport using oil-powered automobiles in Europe. Cars are the ultimate technology for mobility. Mobility means being able to go wherever you want. Vehicles have never been engineered better. The range of affordability and massive infrastructure assets provides unlimited mobility in perfect comfort. Diesel and petrol fuel are not sustainable, and neither are the air pollution, investment in roads and repairs, consumption of land for pavement and parking. Everyone can have a car, but there are way too many cars. Cars and fuel are affordable, but transport poverty is a growing problem and the long commutes to find affordable housing are eroding families quality of life. Fuel is affordable and abundant, but a price shock in oil during a fiscal year induces economic recession. Use of oil in personal cars for transport in Europe is a perfect wicked problem.

One hundred years ago, the internal combustion engine–driven personal vehicle was being manufactured and was affordable only to the wealthy. Even though there were only a few vehicles, they were becoming a management problem for cities. If there was a spike in motorcar fuel price, it did not affect the economy. Over the past 100 years, there has been exponential growth in everything related to personal vehicles. One of the key factors was government price control on fuel. The purchase of the first car for a family would have created a profound increase in mobility compared to not having a car. The OPEC oil embargo and the oil shocks of the 1970s caused a recession and temporary hiatus in the exponential growth.

The 2008 oil shock triggered the global economic recession. Since then oil consumption and personal vehicle travel have been flat and trending downward in Europe. The BAU future scenario has some improved emissions requirements for new vehicles but no real change in infrastructure or personal vehicle mobility. There is a lot of discussion about sustainable transport, but there are objectively no foreseen real disruptions to individual ownership, parking and provision of infrastructure for total mobility. The technology wedge being discussed the most is the battery electric vehicle (BEV). More public investment in private mobility has been made to provide charging stations, subsidies for purchase and exemptions to congestion pricing. The cost of road infrastructure, congestion and other problems are not alleviated by more purchases of BEVs. New problems in grid loads are possible. In 2017, the BEV share of sales in Europe was 1.74% at 307,000 vehicles, up from no BEV sales in 2010 (www.ev-volumes.com). Europe's automakers are bringing more models into the market, but most are plug-in hybrid vehicles. A recent report indicates that carmakers were selling much fewer electric vehicles than their targets in 2016 (Transport and Environment 2017). Several countries have set targets for 100% electric vehicles, including the proposed ban of fossil fuel vehicles by 2050 in the Netherlands and Germany's 100% zero emission vehicles (ZEVs) for new car sales by 2030. The Bloomberg New Energy Outlook puts the most optimistic new car ZEV sales at 22% by 2025. If the number of petrol vehicles remains steady or continues to grow, then the BEV contribution to fossil fuel reduction can be assumed to be about 20%. The purchase of a BEV does not have any effect on mobility and does not provide any new benefits to the vehicle owner or society other than the use of more electricity and less petroleum fuel.

One hundred years from now, oil will not be used for personal transportation and trucks. Coal and gas will not be used for electricity generation. Our path-breaking concept is that the cities and landscape of Europe is essentially unchanged over the ensuing 100 years, but transportation is predominantly by train, tram, electric cycle, electric cart, electric scooter, bicycle and walking. Figure E4.2 shows a street in Munich in 2018 and 2118. The thing that is really different is the arrangement of activity systems within the cities and landscapes. 75% of people's homes are within walking and cycling distance of activities. The other 25% are able to access activities by a combination of public transport, walking and cycling. The historical city streets are sufficient for handling the cycle traffic. There are more extensive tram networks in most cities than 2018. But the most striking things we observe are the repurposing of buildings so that work and market areas are accessible to residences. Home-finding and job locating search engines help employers to co-locate into the communities of their workers. Backcasting shows that, in the past, there were connections between housing, transport and large employment centres prior to the uptake of automobiles. One trigger is the 2008 oil price spike, another is geospatial data availability, and another is the housing affordability crisis in most of Europe.

Our shift project is an insurance product. Insurance provides a hedge against losses. The insurance is an oil price spike hedge. Companies and individuals can purchase insurance to pay 50%–100% of fuel price increases in a given fiscal year. This allows customers to have a known forward price for fuel and reduced exposure to fuel price spikes. Premiums are based on fuel use,

FIGURE E4.2   A street in the city of Munich, Germany, in 2018, and in 2118 in the post-oil era.

so a new specialty of energy auditing of transport develops. The premiums are lower if the customer has a fuel use reduction plan. Government policy starts to require companies and organizations to purchase the insurance to reduce the recessionary risk. The results of the transport energy audits and fuel reduction plans include actionable energy reduction measurement and monitoring opportunities that were not perceived before. The measurement and monitoring also start to reveal the patterns of accessibility and adaptive capacity of transport in relation to choices of location for housing and destinations. As the information grows, new ideas and services emerge that provide services for people to audit their own transport activities; organize their activity systems in the urban form; improve their fitness; and down-shift their dependence on automobiles, the time spent in transport, and the costs of transport.

4.9   DISCUSSION

The fundamental requirement of transition is to leave most of the remaining fossil fuel reserves in place in the geologic formations where they are found. Extraction and use in the future may be possible at much lower levels, once the planetary thermal equilibrium is established and the extent of the resulting climate change is understood. Possibly the most revolutionary projects are the transition of oil, gas and coal companies. The old era of exploring for more oil and gas and developing new production is ending. The new era of exploring for projects that change existing systems to fossil carbon-free ways to meet essential needs is beginning. Transition Engineering can develop at an unprecedented rate and deliver changes faster than any of the previous engineering transitions because we already know the systems that need to change and we have a huge backlog of 'things we must do'. Innovating, engineering and managing the energy transition for households, cities, businesses and communities is not a discussion point; it is the responsibility of every engineer on every job, just like safety and security.

The most important tools for successful InTIME projects are resourcefulness, creativity and a new perspective. Why do we need a new perspective? The Climate Stabilization Wedge is an example of a new visualization of the old perspective that demand growth will continue while carbon emissions will decline by increasing use of green technologies to 'redirect the flows of carbon' (Pacala 2004). Adding more insights into the old perspective is fine, but it does not create a path-breaking concept from the BAU trajectory.

The transition perspective is that demand will decline. We use Transition Engineering tools to help us discover the down-shift perspective while still being able to understand the growth perspective we normally see. In the down-shift era, *not* using energy and materials are relevant options that have more benefits we can discover and develop.

In every InTIME project we have explored, BAU is not a possible future. Investments in projects that assume continued growth in electricity demand or the dominance of private vehicles in developed countries are committed to failure. The technology development that would be required to provide this assumed growth to 2055 is not feasible. The transition pathway represents the reduction from current emissions by redesign and redevelopment of aging infrastructure and technology, and curtailment of expansion in fossil fuel production. We are at an inflection point in history where the historical pattern of increasing risk changes to a path of mitigation of risks.

4.9.1 THE SEVEN WISE MEN AND THE ELEPHANT

The traditional story of the seven blindfolded wise men and the elephant describes the situation of experts in different areas with their own expert understanding of one part of the system but with limited points of view about the whole system. The story involves the failure to put together an understanding of the whole system.

There once was a kingdom that was prosperous and peaceful. The king had seven expert advisors who were in charge of the most important systems: agriculture, water, fuel, commerce, infrastructure, education and health. The generals warned of a looming conflict with a neighbouring kingdom, and the king dedicated a lot of money to raising an army, building a defensive wall and making weapons. This meant the forest was cut down; the agriculture was over-taxed to feed the soldiers; the civic infrastructure couldn't be repaired because of labour and fuel shortages; and the river was polluted by the eroding land, causing sickness. Commerce seemed to be a bit better because of all the new spending, but it didn't really generate new revenues to take care of the problems in education and health. The king called his seven advisors to ask them what was wrong with the farms, forests, rivers, markets, infrastructure and services. Why did it seem like everything was crumbling even as he was spending more money than ever and was actually going into debt?

The king wanted solutions now. The seven advisors argued bitterly about not having the funding they needed and the problems of growing unrest. But they had no solutions for the king. The king lost his patience and ordered his generals to blindfold the seven advisors and put them in a room with an elephant. The king told the advisors that they had to tell him what was in the room before his servant girl came with his lunch, and if she could tell him correctly what was in the room, then he had no need for wise advisors. The seven advisors argued about the evidence – It is a snake. No, it is a rope! Clearly it is a tree. What do you mean, I can feel it is a wall. But it moves just like the king's tent. Are you crazy? It is hard like a battering ram! Each man had a valid but incorrect assessment of what is in the room with them, based on their individual perspective and available data. But, of course, they cannot get an understanding of the whole system by arguing about the parts.

The servant girl comes with lunch and the king asks her to tell the seven wisest men in the kingdom what is in the room with them. The girl answers, 'Of course, it is an elephant, your Highness.' The advisors protest that it was not fair to ask them to solve a problem when they didn't have all the information. The king told them, 'I never ordered that you not take off the blindfolds and look.' They removed their blindfolds and saw the generals and had a good idea of what the real problem in the kingdom might be.

In some versions of this story, the blindfolded men yell at each other and get so angry at not being listened to that they resort to violence. In some versions the men are blind so they don't have the option of removing their blindfolds. I have not found a version where the seven men (it is always men) decide to listen to each other and try to work out the clues. The important lessons are that the collective wisdom of experts with entrenched perspectives is not as accurate as the simple observation about the whole, that it is inherently difficult to put together expertise from different fields, and that changing your perspective by 'lifting your blindfold' and seeing the whole picture is necessary and possibly not that hard to do.

5 InTIME Models and Methods

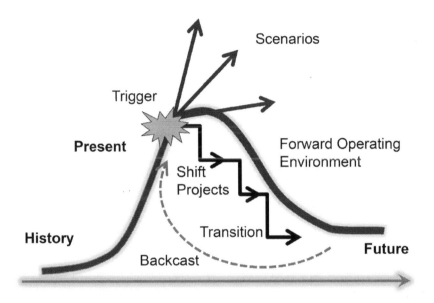

Keep it simple.

The seven-step Transition Engineering method described in Chapter 4 is an effective approach to wicked problems. Working through the different perspectives of past, present, expectations, future under constraints and arriving back again works well for groups to discover new perspectives and to find the triggers for change that become evident from these different perspectives. For multi-disciplinary groups working through the seven-step method, it takes several weeks to gather the data and carry out the analysis, as well as attending workshops for brainstorming. This chapter describes fundamentals and tools that have been developed through research and practice. The field is still developing, and useful ideas are emerging all the time. Professional transition engineers have developed the interdisciplinary transition management, innovation and engineering (InTIME) workflow structure that has proven useful for working with clients in industry to put the down-shift projects into action.

5.1 THE InTIME WORKFLOW STRUCTURE

Figure 5.1 shows the seven steps of the interdisciplinary transition innovation, engineering and management (InTIME) approach organized into three types of transition work: management and operations, innovation and exploration and engineering design and development projects. Transition management of existing systems includes energy auditing and improving efficiency through maintenance and operations. Energy management is an established field in buildings, but energy transition requires growth in the field and broader application in transportation and industry. Innovation has historically been the engine of growth. However, transition innovation is about a down-shift in

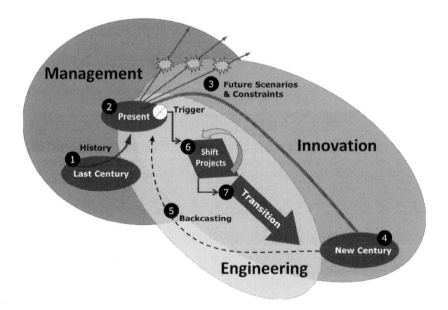

FIGURE 5.1 The InTIME approach illustrated in three groupings of work around innovation, management and engineering projects.

energy and material consumption, realistic assessments of future scenarios and green technologies, and communicating the practical vision of a climate-safe future.

The ideas about future scenarios are explored through conversations with stakeholders, where the results of quantitatively and technically relevant modelling is used to shift expectations. The innovation work requires brainstorming sessions, which need to be carefully planned and executed, with participation by people familiar with the wicked problem and the organization. The engineering work uses the results of Steps 1–5 of the methodology to develop and carry out shift projects, monitor the performance, learn from the early failures and successes, disseminate that learning to the field, improve the field and integrate the new technologies and tools for down-shifting into the whole system. The engineering work includes informing policy, economics and society about how to adapt to the transitions brought about by the down-shift, and facilitating adaptation to better quality of life through the down-shift of unsustainable activities and products.

TRANSITION ENGINEERING CONCEPT 5

If you can't find a way forward, change your perspective.

5.2   FEEDBACK CONTROL THEORY AND ANTHROPOGENIC SYSTEM DYNAMICS

This section presents a dynamic model of human systems using feedback control theory (Krumdieck 2014). Feedback control is required for stable operation of any dynamic system, particularly when there are constraints on the inputs, outputs or operating range. How well a system responds to disturbances and to feedback is a measure of the robustness and reliability of the system. Control system theory is a fundamental representation of dynamic system behaviour, which can be applied, in principle, to mechanical, electrical, biological and ecological systems. The concept of feedback control theory is central to the systems thinking used in the InTIME methodology.

Figure 5.2 shows the well-known feedback control model (Palm 2000). This block diagram representation includes the key functions of a feedback control system according to the standards of the Institute of Electrical and Electronics Engineers (IEEE). With the exception of the external inputs of energy and outputs of emissions, the arrows in the block diagram do not represent flows of materials but instead are communication signals between elements. The control system is an integral part of a continuously operating dynamic system, for example, a cruise control system on a car travelling on a freeway.

The directive represents the motivating input to the system, which is independent of the actual system behaviour but expresses the desired system performance. The reference elements convert the directive into the values of the reference signal for a particular system in a particular situation. The comparator performs the

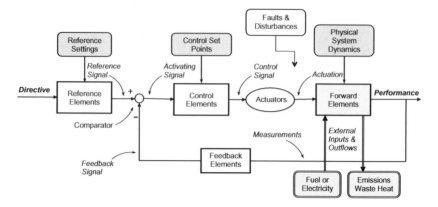

FIGURE 5.2 Block diagram of the feedback control theory model of dynamic electro-mechanical systems.

function of determining if the reference signal is equal to the feedback signal. If the signals are equivalent, then there is no change needed in the system in order to meet the operational directive. The activating signal to the control elements is used to determine operational changes that will push the system performance toward the reference signal values. The control elements generate the control signals according to the magnitude of the activating signal and according to pre-existing control design and set points. Control actuators cause physical changes through built-in physical mechanisms and actuators. Forward elements represent the physical plant and the system dynamics that react to the actuators and affect the performance of the system. The performance is measured by detectors and represents the actual system behaviour. Feedback elements translate measurements of the system performance into the feedback signal, which has the same calibration as the reference signal and thus can be used by the control elements to attenuate the system behaviour.

5.2.1 EXAMPLE FEEDBACK CONTROL FOR AUTOMOBILE CRUISE CONTROL

You may already be familiar with the automatic cruise control on a vehicle. The system directive would be the desired speed set by the driver; the performance would be the actual speed; the set points would be the relationships between controller signals and automobile dynamics; the actuator would be the fuel supply throttle and brake fluid; the external inputs would be fuel and air; and the forward elements would be the entire vehicle, including the seat for the driver, engine thermodynamics, transmission, aerodynamics and the friction between the tires and the road.

The driver sets the directive of a desired speed according to the posted speed limit and the present driving conditions. That speed in the mind of the driver cannot be used by the microprocessor in the car; rather, it is first translated from a choice to a set point by pushing a button, then processed into an electronic reference signal. The measurement of vehicle speed is achieved through a transducer that produces an electronic signal that would not be used by the cruise control processor or recognized by the driver as speed unless that signal were to be processed through calibrated electronics, such as a Wheatstone bridge, and then sent to a speedometer to actuate the indicating needle. Another important concept is that the controller acts in predictable ways in response to the control signal, which is not the actual speed but the difference between the reference and the feedback signal.

The cruise control system for a given automobile was designed and calibrated, and the microprocessor controller was programmed according to the automobile specifications before the driver got into the car, and the driver did not participate in any of that design or manufacturing. It is vitally important to understand that the controller can only actuate the system according to the existing design and cannot operate the system in a way that changes the design. Unsafe behaviour can definitely crash the car, representing a system failure outside the capabilities of the cruise controller. The driver cannot change the environmental impacts already caused by manufacturing the car, and the driver cannot change the fuel used by the car or other things about the physical nature of the car, like the number of cylinders or the fuel efficiency. The driver could choose a lower set-point speed, which would reduce aerodynamic drag and improve the fuel efficiency of the vehicle to a certain extent. The driver could also choose to take a train or ride a bike, which would down-shift the emissions for the trip.

5.2.2 FEEDBACK CONTROL WITH HUMANS AS DIFFUSE CONTROL

The main implication of the feedback control theory representation of anthropogenic systems is that changing behaviour is not an effective way to achieve sustainability. Behaviour will quickly adapt to changes in the physical systems, but behaviour cannot change to use the physical systems in ways that do not work. Figure 5.3 shows

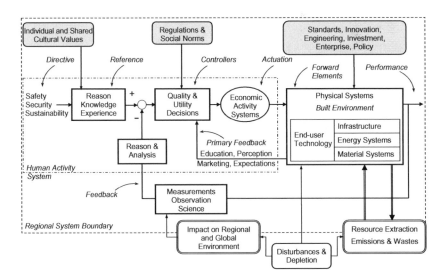

FIGURE 5.3 Feedback control theory model of the human activity system and the energy system.

the feedback control theory model applied to human activities. All of the control elements are collective human processes. The signal processors, shown as squares in Figure 5.3, are cognitive frames of reference. Throughout human history, when change in technology and society was very slow, the cognitive frames would have been traditions.

The overall directive for individuals, groups and societies is *survival*. The particular aspects of survival at different scales are safety, security and sustainability. People want to be safe and secure, so they follow the traditional practices of their culture, make choices within a prescribed decision set, and receive reinforcement of the success of their way of life through the feedback of continuity. If things work according to expectations, then the traditional culture continues. Sustainability has been achieved by having traditional values, behaviour and resource consumption that were in balance with availability. The future for traditional societies would be like the past because the collective decisions and the regulations from leader groups ensured the continuation of the past systems and behaviours. Regulations and enforcement of social norms through various means act as set points on the diffuse control in the anthropogenic system.

The energy system is part of the physical system, which includes infrastructure, the built environment and technologies. The energy system includes the physical systems, with the inflow of extracted resources, conversion and use, and the flow of combustion products back into the environment. The open flow model is a reasonable model for energy engineers when their job involves keeping the system growing and flowing. Open flow control is used in systems like vehicle fuel supply, where the end-use demand is supplied from a large storage tank. However, if the end-use grows over time, the limits of the system operation can be reached. At that point, it may be more cost-effective to develop ways to stop demand growth than to build a larger supply system. If the end-use keeps growing, the limits of the resource may be reached, at which point, control of the end-use is the only option. The Transition Engineering approach proposes that recognizing resource limits, and using that information to develop projects that change the system and down-shift demand, is more likely to provide the conditions for economic prosperity than continuing with projects that assume sustained growth. Simply stated, it is preferable to transition to stable operation than to have booming growth and then go bust.

The single thickness arrows are not flows of energy; they are signals, decisions, relationships and actions. The activity systems consume resources and produce pollution and other impacts, but the performance is actually the quality of life achieved by all of the activities and how sustainable that quality of life is. The standard of living is a well-known measure that includes factors like literacy, life expectancy, nutrition and childhood health as well as energy and material consumption. It is possible to have a low quality of life while having a high standard of living. It is also possible to have a high quality of life while having low energy and material consumption. With a new perspective we can develop measures of system performance that do not rely on consumption and monetary transactions but rather measure the quality of life achieved by the activity system. The system performance is determined primarily by the physical system characteristics and, to a lesser degree, by behaviour. People can only use the appliances and energy products that are available in the market, and they typically use appliances, vehicles and buildings the way they were designed; according to the instructions; and within the range allowed by cultural norms, rules and regulations.

The actuators in the system are the market interactions to purchase the energy to meet the loads caused by activities arising from normal behaviour. Manufacturers and developers make changes in the system in order to improve profit or to improve performance. Feedback about safety and security failures can also cause changes. For example, safety issues for utility line workers, electricians and homeowners led to development of standards in worker protection gear, wiring and appliances to reduce hazards. When pollution from factories and cars caused unacceptable health issues and environmental damage, electrostatic precipitators, wet scrubbers and catalytic converters were developed. However, these corrective changes to existing systems have historically taken long periods.

The primary feedback in the anthropogenic system is the experience of what works. People use their experience to decide what to do next. People are most likely to choose to use energy tomorrow in the same way that they did today if it worked, which is only logical. No one in Vancouver can choose to use a hydrogen car tomorrow because there are no hydrogen cars. No one in Vienna can choose to take a horse and buggy into town today, even though it would have been the norm in 1880. Commuters in San Francisco are amongst the few people in the United States who can still choose to take a tram, because such systems were dismantled in all other cities. The behaviour of how people use energy and how they choose energy-consuming appliances is primarily determined by the physical system.

Scientists can measure secondary feedback about the unsustainable impacts resulting from normal operation of the physical system. However, the signal must have direct relation to the decision set of each person and organization in order to influence behaviour. And, as just stated, the physical system actually determines behaviour through the primary feedback of habit. There is one key conclusion from this examination of the anthropogenic system: engineers and developers, under the pressure of policy and in a manner that responds to the market, must change the engineered physical system to mitigate the impacts of the mega-issues of climate disruption and oil and resource depletion.

Prices, policies, social values and especially the hundreds of thousands of individual choices cannot change the energy system; energy engineers must deliver the energy transition, and then the prices, policies and social behaviour will adapt to use the new system. This is an important finding from examining the anthropogenic activity system using the feedback control model. Engineers cannot simply accept the narrative that climate action or sustainability will be achieved through political leadership, a carbon tax or consumer behaviour change. Engineers must get to work on the job of serving the interests of society, and we need to do it with new tools InTIME.

5.3   DEVELOPMENT VECTOR ANALYSIS

The development vector analysis approach looks at the probability that a technology will penetrate the market within the target timeframe. For every new technology idea, there are different feasibility hurdles that must be overcome along the development

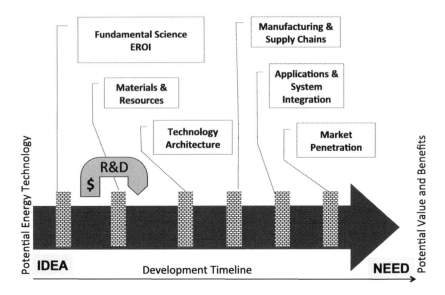

FIGURE 5.4 Technology development vector analysis showing the hurdles a new technology must overcome to move from an idea to a product that meets a need. (Reprinted from Springer Nature: *Transition Engineering* by Krumdieck 2017. With permission.)

pathway. Figure 5.4 illustrates the development pathway for an idea that is meant to meet a need in the activity system. The figure sets out the barriers to realization, and the hurdles that must be overcome with investment in research and development (R&D).

The first step in the analysis is to clearly define the need and how the idea, if it could be successfully developed and marketed, would meet the need. The first question in product development is, 'What is the need for this?'. To accomplish the energy transition, fossil fuel needs to remain in the ground. How does the idea reduce the number of coal-fired power plants, kilometres travelled or commercial flights? It is easy to lose track of the ultimate need to significantly reduce fossil fuel production. For example, in 2016 a new idea of self-driving cars (SDCs), also known as autonomous vehicles, became very popular. The self-driving cars depend on sensors, Global Positioning Systems (GPS) and algorithms to navigate. By 2017, media coverage of the idea was pervasive; Audi, Ford and other automakers announced their plans to launch vehicles in 2020, and governments were trying to figure out what the road rules would be for SDCs. What is the need for SDCs? What are the benefits and the value to society now and 100 years from now?

5.3.1 THE FIRST HURDLE: FUNDAMENTAL LAWS OF PHYSICS AND BIOPHYSICAL ECONOMICS

The fundamental laws of science, such as the first and second laws of thermodynamics, cannot be violated, regardless of how exciting the idea is. An example of the first fundamental science hurdle is the geology for stable gas sequestration for the hundreds of years needed for the idea of coal-fired power generation with carbon capture and sequestration (CCS). The vast majority of gas that has formed over hundreds of millions of years in sediments has migrated through rocks and soil and escaped into the atmosphere. The gas reservoirs that are currently being exploited represent very rare circumstances where gases were trapped. The idea is to put waste gases into exhausted geologic features that held gas before. There are other EOR CO_2 injection projects, but only this one purports to be a permanent storage. It also is actually maintaining gas production pressure in the reservoirs. To date, geological science is clear that only very special geological conditions can be expected to contain injected gas. The energy return on energy invested (EROI) is a fundamental hurdle for an energy production technology, particularly if it is less than 1, as in the case of some biofuels, or negative net energy, as is the case for batteries and hydrogen.

5.3.2   THE SECOND HURDLE: MATERIALS AND RESOURCE AVAILABILITY

Many of the new energy technology ideas require development of new materials and new material processing like electrolytes and anodes for batteries and catalysts for fuel cells. Chapter 2 discussed some of the key materials for energy systems. Copper is obviously the primary energy material as it conducts all of the electricity we use. Scarcity of minerals represents a hurdle to growth in development. Materials science and engineering of new or advanced materials is also a barrier. For example, trapping hydrogen in fullerene (C_{60}) is an idea for hydrogen energy storage. The hurdle is that hydrogen can penetrate any solid material, so how would you make a containment vessel? Materials scientists have proposed making materials that have very open nanostructures that could adsorb and desorb hydrogen. Fullerene has been proposed, but the adsorption isn't high enough, so a new idea about coating the fullerene with a monolayer of calcium is researched by numerical modelling (Yoon et al. 2008). However, even though there are thousands of research papers on fullerenes and hydrogen, there has been no fullerene material that has cleared the second hurdle to be used in a technology.

5.3.3   THE THIRD HURDLE: TECHNOLOGY EMBODIMENT AND SYSTEM ARCHITECTURE

The architecture refers to how the technology will be realized, for example, thin film or crystalline solar PV or algae farming in open ponds or glass tubes. The architecture hurdle can also involve the processes necessary for the technology to work. For example, using hydrogen as an energy carrier would not be possible if it is not compressed to at least 800 bar and stored in a safe container (Bossel 2003). The Stirling engine is an example of a technology that has a theoretically possible thermodynamic cycle, and it does not require any materials that do not yet exist. The technology has been embodied in numerous different architectures with one to four pistons, and horizontal and vertical orientation. However, few Stirling engines have made it over this hurdle.

5.3.4 The Valley of Death Hurdles: Manufacturing and System Integration

A new technology will require manufacturing; integration with existing energy and end-use systems, including regulation; and marketing at an acceptable price. For example, biofuels can only be used in blends of less than 10% in many automobiles. Another integration example is electric vehicles, which have limited use without charging stations being available at points along a route. The time for such integration infrastructure and technology change to develop would necessarily limit the market penetration of the technology. Market penetration depends on cost, but also on need, regulation, perception, marketing and confidence. For example, a few Stirling engines have been developed to the point of market availability. The WhisperGen Stirling engine was developed in New Zealand with an architecture using four pistons and using natural gas for fuel. The system integration strategy is to use the unit to provide domestic hot water, and during water heating on demand, the unit would also produce up to 1 kW of electricity, which could be used in the household (Stuff 2009).

5.3.5 Technology Readiness and Market Penetration

The implementation hurdles for new technologies can be anticipated by observing how previous technologies moved into the market. The level of technology readiness is usually rated on a scale, as follows:

1. Innovative idea for a way to meet a need
3. Laboratory demonstrations of the principles
5. Conceptual prototype demonstration
7. Functional prototype demonstration
9. Manufacturing and supply chain established, and offering into a test market
10. Achieve market share and become competitive

R&D into hydrogen and fuel cells was popular in the early 2000s, and many articles in popular science media predicted uptake in the market within the decade. Biofuels gained renewed interest as oil prices started to climb above $50/barrel from 2005. The United States and Europe an Union instituted policies requiring fuel retailers to supply certain percentages of biofuel. First-generation biofuels processed from food crops were quickly understood to be problematic, and R&D efforts focused on second-generation liquid fuel from wood and third-generation fuel from algae. As of 2015, only ethanol and biodiesel from crops are commercially available, and US ethanol production reached a plateau around 2009 (RFA 2015). Electric cars have a popular image as a substitute for petroleum fuels, but as of 2015, the outlook for electric cars is no better than that of hybrid cars, which have failed to increase market share above 5%. In 2015 the US take-rate for alternative vehicles was negligible despite tax credits and incentives: 2.16% for all petrol hybrids, 0.21% for plug-in hybrids and 0.35% for battery electric vehicles (BEVs) (Cobb 2015).

Between 2004 and 2011, the total investment in renewable energy grew from $40 billion to a high of $279 billion, and remained over $200 billion, but it has not continued growing at the same rate as government incentives have started to be reduced (Frankfurt 2015). The largest share of global renewable energy still comes from traditional biomass, for example, wood and dung (13%), for heating and cooking, and much of the growth in biomass has been corn ethanol and biodiesel from canola and palm kernel oil. Hydropower remains the largest renewable share of global energy used for electricity generation (3.2%). Most of the commercial investment in renewable electricity generation has been in wind farms and solar PV panels. Despite being the largest potential baseload renewable resource, geothermal energy provides less than 1% of global electricity capacity, largely due to the development risks, even though the technology is mature (Ram 2015).

5.4   STRATEGIC ANALYSIS OF COMPLEX SYSTEMS

The strategic analysis of complex systems (SACS) approach is a tool that can be used for the 100-year concept generation, backcasting and the shift project development processes (Krumdieck 2009). The SACS method helps the engineering team organize the complex development issues into a set of modelling investigations. It also has been found to be a very effective communication tool with all of the stakeholders, including council workers, political leaders, businesses and members of the public. The procedure for the SACS approach is set out in Figure 5.5, and the four steps are described below. The SACS method has been used for several large projects, such as long-term planning for the city of Dunedin, New Zealand (Krumdieck 2010).

5.4.1   CHARACTERIZE THE ACTIVITY SYSTEM AND ENERGY RESOURCES (INTIME STEPS 1–3)

The first procedure in the strategic analysis is to define the activity system and carry out the first three investigations of the InTIME methodology (e.g. gather data about the present situation, study the history and explore future scenarios). The activity system will necessarily involve a community, organization or company. The InTIME methodology requires working with stakeholders who participate in the activity system or who are affected by it. Engineers may have limited experience working with communities and stakeholders. There are discussion and workshop methods that can be used, including the *TransitionScape* workshop approach, which has been demonstrated for InTIME projects with communities who are concerned with the mega-issues of unsustainability (Krumdieck 2012). Define the activity system and the key energy resources to be studied for transition. Use Geographic Information System (GIS) data, (freely available through Google map for example) to develop a geographical map of the activity system. Understand the population, culture and society. Make an inventory of the infrastructure currently used for the activity system, and estimate the condition, maintenance and remaining usable life. Make an inventory of the resources needed for the activity system such as energy, water, land and materials, and estimate the quantities used and the efficiency of use. Define the timeframe for analysis.

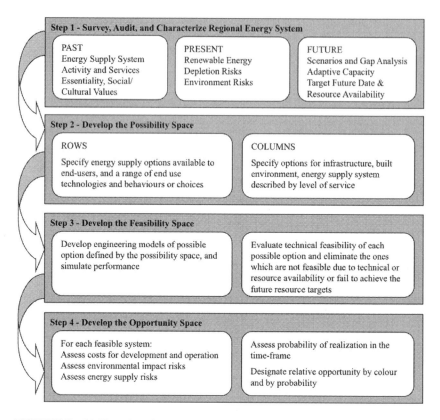

FIGURE 5.5   SACS methodology for generating path-breaking concepts.

Research the history of the place, the people, and how they carried out the activity 100 years ago. Hold discussions with different stakeholders about what they think the problems with unsustainability are and to propose ideas for solutions. At this stage, do not analyze or give opinions on the solutions that people believe might be possible. The results of the future scenario analysis and the forward operating environment (FOE) will be presented at a later stage to the community as part of the proposal for the shift project. Establish the FOE for the constrained resources. Use resource data, depletion curves and information from scientists to model the resource availability envelope for the critical resource in question at the target future date.

5.4.2   CREATE THE POSSIBILITY SPACE

For Step 2 in the SACS methodology, a 'concept' means the 'things we could do' when brainstorming shift project ideas, and means 'ways things could be' when brainstorming 100-year concepts. Figure 5.6 gives the general layout for the possibility space. Arrange the concepts to form the possibility space by separating the concepts according to whether they involve the built environment or end-user choices.

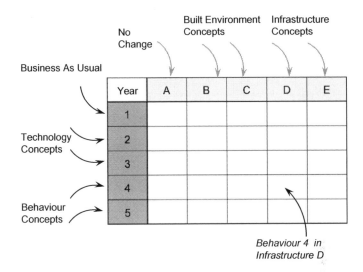

FIGURE 5.6 Matrix form of the possibility space generated by the SACS method.

Columns = built environment and infrastructure options
Rows = end-use technologies, resources or behaviour change options.

The matrix format is used to organize the concepts and to define the systems for quantitative analysis. The matrix is also used to communicate the results to all stakeholders in a way that includes all of the ideas and concepts brought forward for consideration but makes it clear which options are possible and which are actually good opportunities. The stakeholders should agree on a future year for the strategic analysis. The ideas that stakeholders provided in Step 1 of the SACS should be included in the analysis. It is usually most helpful to reserve the first column for the current infrastructure and the first row for the current energy technologies and behaviours if there were no changes in the current system (e.g. a no-growth scenario) or if there were a continuous growth along the business-as-usual (BAU) path to the year of the analysis.

The energy activity system operates within the context of public and private investments. The built environment refers to buildings, including homes, schools, offices and shops. Built environment is usually private property and involves the decisions of what to build where, but the city zoning and land use planning affect these market decisions. Infrastructure refers to roads, railways, water supply, electricity networks and waste management. Governments usually build infrastructure with public money, but the need for infrastructure is determined by the current or planned private property developments. Technologies are things that can be used in the built environment or in the infrastructure. Households own the appliances in their homes, but government regulation sets efficiency and size standards. Freight companies own trucks, trains and airplanes, but governments usually build roads, rail lines and airports.

Behaviour refers to the choices of where to live, what transport mode to choose, what size house to purchase or how many refrigerators to own. The technology is used and the choices that constitute behaviour are made in the settings of the infrastructure and the built environment. For example, in transportation systems, options in columns would be associated with the urban form, for example, transit systems, cycle and walking infrastructure, and new types of developments within the existing urban form like apartment buildings or eco-villages. Rows would be technologies like high-efficiency vehicles, electric cycles, and alternative fuels, and also behaviour like travel mode and choices to live near destinations (Krumdieck 2011). The options are best described by a percentage change from the present activity system. For example, consider a SACS timeframe of 2030. A built environment concept for office buildings could be a net zero energy retrofit of 25% of existing buildings. An infrastructure concept for a central city area could be increasing the conversion to pedestrian-only streets from 1% to 20%. A technology option could be increased share of electric vehicles from the present 0.01% to 25%. A behaviour concept could be reduction of the penetration of appliances like electric clothes dryers from 85% to 10%.

Characterize the performance and energy use for each of the infrastructure development concepts with the particular technology or behaviour adaptations. Use high-level modelling of the change in fossil energy use in the target year. Make sure to characterize the options in terms of the specific local system and the degree of change over the time period of the analysis. For example, reduction in kilometres driven per year per vehicle by 25% will result in a reduction in gasoline consumption by 25%.

5.4.3 Generate the Feasibility Space

Assess the technical and resource feasibility of each opportunity combination to form the feasibility space. Technical feasibility and resource availability are the criteria for Step 3; at this point, do not consider cost or whether policies or market acceptance exist to support the options at the current time. The feasibility for each infrastructure or technology option is determined by physical constraints. The feasibility depends on how the calculated result fits with the constraints set for the analysis. If the combination is not feasible, then the cell is blacked out, as shown in Figure 5.7.

Assess the development potential by estimating the likelihood that each of the concepts could be developed by the target date of the analysis. This likelihood of availability can be influenced by expected depletion of key resources, available supply chain and manufacturing, competition with lower-cost options, government policy, replacement rate, and so on. Rank each feasible possibility as:

1. *Likely*: the option is currently available with good performance, there is no constraint on production or distribution, policy supports uptake, and market growth could meet the target.
2. *Possible*: the option is currently feasible with demonstrated performance, constraints and policy support could be put in place, market is currently small.

	Year	Built Environment Concepts		Infrastructure Concepts	
		A	B	C	D	E
	1	▨	No	Possible	No	No
	2	Possible	▨	Possible	No	▨
Technology Concepts	3	Possible	No	Possible	Unlikely	▨
Behaviour Concepts	4	Unlikely	No	Likely	Likely	▨
	5	Unlikely	No	Likely	No	▨

FIGURE 5.7 Feasibility space showing how options that do not meet the criteria set for the strategic analysis are deemed unfeasible and are not studied further. The relative likelihood of a combination of infrastructure and behaviour concepts being realized by the target date is also estimated.

3. *Unlikely*: the option is currently at the research stage with no demonstrated performance and does not meet the needs. Constraints in production or delivery are present and not yet overcome. Policy support is not expected. There is no current market and limited future market. The EROI is less than 10.

4. *No Possibility*: the option is not demonstrated at the research scale. The option has one or more significant development hurdles. The option is not better than other possible options, it requires more resources and more policy support than the benefits derived. The EROI is less than 5.

For example, consider the residential electricity use for Bristol, United Kingdom, where the average annual electricity use per person is currently 5,130 kWh, and the population is 450,000. An infrastructure option is to supply all residential electricity by roof-mounted solar PV. The Bristol residential energy demand is 2,309 GWh. The total land area is 106,000,000 m^2, of which roughly one-third is rooftops and the solar PV resource is about 140 W/m^2, which means that the total solar PV generation would be 4.9 GWh. The BAU behaviour option is clearly not feasible. Another behaviour option is that 100% of residents would be connected to the smart grid, and allowing time-of-use energy management and efficient behaviour reduces electricity demand per person by half. This combination of running 100% of Bristol's residential electricity demand from rooftop solar PV is not feasible even with a 50% demand reduction in 2050, because it is not physically possible. These option combinations would be blacked out in the possibility space. Now consider a built environment option that, by 2050, 50% of residential units are retrofit to low-energy standard that reduces electricity demand to 1,000 kWh/yr. Research shows that the technologies are feasible for current electricity end-uses, government policy supports insulation and other retrofits, and there are many older buildings in high-value locations that could be retrofit to new standards due to market forces. Thus, we would rate this built environment option as 'Likely'.

5.4.4 GENERATE THE OPPORTUNITY SPACE

The final Step 4 in the SACS analysis evaluates risk, viability, prosperity and costs of the concept combinations. The resulting opportunity space will be used to communicate with stakeholders about the positive opportunities for infrastructure investments, supporting built environment developments, developing policies for subsidizing technologies and regulations for behaviour changes to meet the transition goals. The matrix analysis presents a clear graphical representation of complex interactions between behaviour and environment and how infrastructure options interact or synergize with the technology options.

Assess the levelized cost of energy (LCOE) with discount rate of 0% for each of the energy supply technologies in the context of the infrastructure concepts. Assess the relative costs of each of the infrastructure changes and the improved energy intensity from behaviour changes. Include as much of the external costs and follow-up costs as possible. The analysis is comparative, so costs do not need to be exact; rather, rank the different options as 'high' 'mid-range' or 'low' cost. A number of dollar signs can be used in the matrix to designate the relative costs. Rank the EROI of each energy supply option. Rank the energy efficiency of each end-use technology, infrastructure and built environment option.

This next step in the analysis can be structured like any other concept evaluation. Different criteria and measures of merit can be established, and the relative score of each remaining option can be determined from the composite score. Often at the concept phase, costs are difficult to quantify, and EROI can vary greatly with small changes in assumptions about lifetime and utilization factor.

Assess the relative risks. Environmental, social, operational or other risks do not have to be quantitative. For example, the operational risk of a solar PV system is higher for a residential system with battery storage and inverter than if the system can be used directly on a commercial site for dedicated DC motors driving ventilation fans. Rank the feasible concepts and use the following colours to designate the overall risk level:

Green: High EROI (>15), low risk, renewable resources, recyclable, long life, enhances learning, improves community, improves health

Yellow: High EROI (>10), moderate risk, limited use of finite resources, reusable, no negative social impacts

Orange: Modest EROI (<5), moderate risk, requires finite resources, negative social impacts, limited recyclability

Red: Low EROI (<2), high risk, requires finite resources, does not improve health, negative correlation with social values

Figure 5.8 shows an example with the composite rankings for cost, EROI, environmental risk, social impacts and further reliance on finite resources indicated by colour. In Figure 5.8, green is the lightest and red is the darkest gray scale, the change in the infrastructure or built environment option C appears to be the best set of opportunities. Infrastructure concept D is a pretty good opportunity as well, but the technology concept 1 is not a good investment under any circumstances.

	Year	A	B	C	D	E
		BAU	**Built Environment Concepts**		**Infrastructure Concepts**	
BAU	1		No $$$	Possible $$	No $$$$	No $$$$
Technology Concepts	2	Possible $		Possible $$	No $$	
	3	Possible $$	No $$$$	Possible $	Unlikely $$	
Behaviour Concepts	4	Unlikely $$$	No $$$$	Likely $	Likely $	
	5	Unlikely $$$	No $$$$	Likely $	No $$$	

FIGURE 5.8 The opportunity space indicating that five of the options, A-2, C-3, C-4, C-5 and D-4, are good opportunities, and that options B and E do not present any transition development opportunities.

Infrastructure concepts C and D could possibly be developed together. None of the technology developments are opportunities for meeting the transition development goals under the infrastructure option E. Option E would be a very bad investment. An example of such a situation might be continued urban sprawl development that makes reducing fossil energy use by any means nearly impossible.

In the transportation example, the costs of some of the options like electric bus and bicycle paths were actually cost savings as the options reduced fuel use over 40 years and paid back the investment cost several times over. Other options, like developing urban village centres, which concentrate destination activities in certain areas around transit hubs, had positive economic development and social benefits. Increased walking and cycling were one of the behaviour options modelled, and this option provided reduced fuel and vehicle costs as well as health benefits.

5.5 THE MATRIX GAME

The language used around the mega-issue of climate change is often militaristic. People speak of saving the planet, as if we are under attack by an invading force. We say we must win the war on the climate and fight for survival. The InTIME methodology borrows the concept of the FOE from military thinking. Defining the FOE means dealing with the real facts and considering all of the information. There are many examples of battles that were won because of one piece of information, one bit of intelligence. A military corps never has the luxury of knowing everything, but its members need to know everything they can. The job of winning a battle hinges on using all of the intelligence, even if it is not positive, and dealing with the facts.

There are examples in history of one side being outnumbered, or with their backs against the wall, and then somehow turning the tide and winning against the odds. Given the facts about the mega-issues, the situation is that we are losing the battle to save the planet and we are on our way to defeat in the war against climate change. Winning against the odds is our favourite Hollywood movie script. The story usually involves a brave hero who takes on an impossible mission, or has a surprising and unexpected idea. Military history often points to these surprising ideas actually being the most obvious solution. For example, during World War II, a German installation on Riva Ridge in Italy was thought to be impregnable because it was on the top of a 600-metre sheer cliff with only one heavily defended approach. It was winter and any approaching soldiers would be easily spotted and fired upon. The US Army 10th Mountain Division captured Riva Ridge without a shot fired because they did the most impossible, most unexpected, but most direct thing – they climbed the cliff under the cover of darkness. Another example is the Battle of Stirling Bridge during the First War of Scottish Independence in 1297. The English force greatly outnumbered the Scots; had better armaments; and, most importantly, had a large cavalry. Soldiers on horseback have an overwhelming advantage over foot soldiers. The Scottish leader, William Wallace, devised a surprising plan to counteract the advantage of the horses by fighting the horses directly. The Scots waited until a substantial number of knights on horseback had crossed the narrow bridge. Then the Scottish spearmen rushed into close quarters, negating the use of the Welsh bowmen across the river, and formed a tight ball formation with the spears sticking out. Wallace had observed that horses charging into a thicket could not see the hazard directly in front of them. Thus, as the knights charged the Scots, the horses were impaled on the spears and the fallen knights were slaughtered, with only one surviving and escaping back across the bridge. The dead horses formed an effective barrier to the foot soldiers.

In order to win the battle for the climate-safe future, what has to happen? What is the surprising move that would turn the tide and follow the trajectory of dramatically reduced emissions? Clearly, the obvious yet surprising answer is – leave the fossil carbon in the ground. How could that possibly happen? The oil and gas companies are so powerful and control so much wealth that they can ensure that production continues apace, and we lose. Wouldn't reduced coal, oil and gas production cause an economic meltdown, as happened during the OPEC oil embargo in the 1970s? Governments are comfortable supporting electric vehicles and solar PV panels, but it would be inconceivable that they would regulate reduced hydrocarbon production directly. Just like climbing the Riva Ridge cliff in the dark, or attacking the horses of the knights straight-on, we have to think about the power of the oil companies, and the value of using the oil, in order to see the answer: The hydrocarbon production companies would reduce production if they made more money doing that than maintaining or increasing production.

5.5.1 Playing the Matrix Game

Military strategists use the matrix game to develop scenarios and action plans. The important premise of the matrix game is that actions and solutions cannot be assumed at the outset. The rules of the game are simple. First, the objective of the game must be defined. What is meant by 'winning'? There can be only one objective and it must be clear. The game has no other purpose or outcome than achieving the objective. The game is played in rounds until the objective is achieved.

Figure 5.9 illustrates the matrix game negotiation platform. Assign the role of the game master to a participant who is the objective director of the game activities. The actors or participating parties with a stake in the game are identified. The actors can be producers, end-users and governing organizations. Assign players to take the roles of each actor. The arguments of the actors must be researched by the players to identify their purpose, their objects, and the ways they operate. The first round of negotiation proceeds by the game master proposing the objective to each actor in turn, while the others listen but cannot interrupt. The actor is asked, if the objective is achieved, what would they do or what would be required from his or her perspective according to their arguments? Each actor proposes negotiations that would allow them to achieve the objective. If all actors agree that they could follow their organizational arguments and achieve the objective, then the game is finished. If the requirements of one or more actors affect other actors, then a second round of negotiation is carried out to achieve the objective with those requirements included. Once all actors agree that their arguments are satisfied and the objective is achievable, then the work of Transition Engineering can be initiated to meet the requirements discovered by the negotiation process.

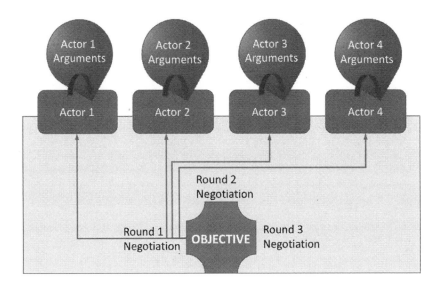

FIGURE 5.9 The matrix game structure. The game is carried out by participants assuming the roles of actors in the system and engaging in negotiation according to the arguments of the actors if the objective were achieved.

5.6 DISCUSSION

Transition Engineering project work has led to the development of the analysis tools described in this chapter. As the field grows and more types of projects are done, new ideas will emerge and the analysis methods will improve. This is the way that all engineering fields evolve. Engineers take on the projects of making things work, try new things, share their discoveries with their peers and develop standards to embody the best practice. Engineering capabilities are developed by practice and learned by dissemination. The Global Association for Transition Engineering (GATE) has been established as the peer professional organization for the field (www.transitionengineering. org). The membership is open to all engineering disciplines and to any other professionals. As we have seen, the InTIME discovery and down-shift projects need innovation and management with an interdisciplinary perspective. Scientists, lawyers, artists, businesspeople, policy developers, social entrepreneurs and economists are all needed to provide expertise and facilitate changes. All of these other disciplines also need a forum where the technical experts can be honest about what is possible and what makes sense. At this early stage of the field, and with this being the first text, the best advice to learn about Transition Engineering is to contact GATE and ask about training, find a consultant or a mentor, or to get involved in ongoing down-shift projects.

Example 5.1 Transition of Burnside, Christchurch, New Zealand

The first 100-year path-breaking brainstorm session was in 2003. The team had two academics, one with expertise in energy systems and sustainability and one with expertise in transportation; two PhD students in mechanical engineering; and two undergraduate students in environmental and landscape design. Additional participants included academics in anthropology, environmental science, geography and agriculture. Brainstorming sessions of 2–4 hours were carried out each week for 10 weeks. Research and modelling were undertaken to explore the ideas developed in the brainstorming sessions and report back to the team on the findings.

The essential activity system was urban activities and the specific location was the Burnside High School (BHS) Zone within the city of Christchurch. Anthropologically, a group of people who agree on the content of the education of the next generation and carry out that education together form the basis for a 'community'. Modern cities have this underlying structure of residential areas forming the zones for schools. Smaller rural towns are models for the way the school forms a central point of connection for the community. The BHS Zone has about 10,000 people, and a geographical radius of 3 km. It takes about 10 minutes to bicycle or 40 minutes to walk from one side to the other. Burnside is a suburb positioned midway between the city centre and urban boundary. GoogleMap™ was not invented at that time, so the team walked and cycled the BHS Zone to evaluate accessibility from residences to schools, shops, churches, medical clinics and places of employment. Accessibility is a measure of travel time by different modes.

> *History:* In 1903, Christchurch was a well-developed European-style city founded in 1850. The BHS Zone was farmland. In the 1950s, the census reported that the main transport mode was cycling for trips in the

city, and an extensive network of electric trams was being removed. The Burnside area was developed as a planned suburb in the 1960s, with the farm lanes largely being retained as the main suburban roads. Construction was started on BHS in 1959. The suburban area was built by the 1970s and has remained basically the same since.

Present: New Zealand has higher car ownership rates than the United States, and 95% of all current trips are by car, according to the NZ Travel Survey. The city of Christchurch has a population of 330,000 people, nearly all in low-density suburban single-family housing with lot sizes greater than 400 m². The flat topography is favourable for cycling, but traffic congestion is becoming a serious issue.

Future: Figure E5.1 shows a tabletop 1:100 scale model of the Burnside zone that was used to measure distances and tabulate land use and activities. The government in 2003 planned for continued growth and had a policy supporting biofuels. To test the biofuel scenario, we carried out a retro-analysis study of the feedstocks for biofuels, including corn, agriculture wastes and canola (rapeseed) in the year 2004 (Krumdieck and Page 2013). The EROI for NZ-grown corn ethanol was 1.56, and for rapeseed biodiesel it was 2.13. The total yield if all corn crops were used as feedstock would be 0.4% of the petrol demand, and if all rapeseed were converted to biodiesel, it would be 0.04% of the diesel demand. Thus, the fuel-biofuel substitution wedge is negligible.

We carried out the probability vector analysis of electric cars to evaluate the idea that the current fleet of petrol vehicles would be 'substituted' to any given extent with electric cars, as shown in Figure E5.2. The current vehicles are largely

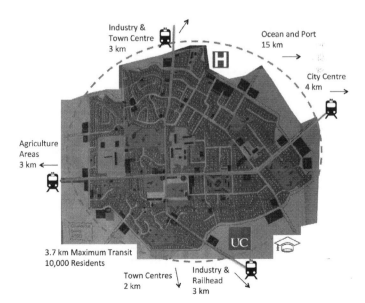

FIGURE E5.1 The path-breaking 100-year time travel to Christchurch was aided by creating a 1:100 scale map and using it to measure travel distances.

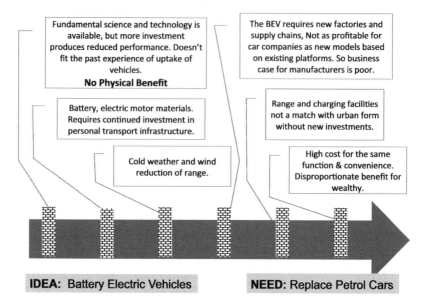

FIGURE E5.2   The technology development vector analysis of BEVs showed that the probability was much higher that people would change the need to travel by vehicle rather than importing more vehicles.

low-cost, second-hand vehicles imported from Japan. The Japanese government supports the car industry by placing high taxes on vehicles older than 5 years, and the market in New Zealand is basically used as a disposal for these older surplus vehicles. The technology exists to manufacture personal electric vehicles (EVs) that 'substitute' to a certain degree for petrol vehicles in Japan. However, the probability that over the next 100 years the same production levels for EVs would exist in Japan that would provide for a low-cost outflow of surplus vehicles to New Zealand was estimated to be low. The anticipated EV supply in 100 years is shared golf cart type vehicles for moving heavier loads and disabled people, electric cycles and scooters, and electric trams and trains. There are also shared trade vehicles being dispatched from trade hubs that do repairs and services. Most of this vehicle manufacture could be done in New Zealand. Thus, our future scenario for personal transport technology has a nearly total shift away from personal vehicles. This scenario leads to numerous benefits in the city, freeing up nearly one-third of the city land to redevelopment, reducing injuries and fatalities, and making cycling and walking much easier, and reducing the transit distance as people no longer have to navigate their way through the vehicle spaces.

The FOE for New Zealand was evaluated to be 100% decline of oil supply for personal, freight and public transport by the end of the century. An 80% reduction in coal is indicated, and the natural gas reserves will have been completely depleted. Thus, the 100-year brainstorming session was about a much-reduced processing sector, renewable electricity, and fossil fuel–free transportation. That seemed like another world.

What things that exist today would still remain in 100 years for the same population? The old, substandard stock of houses, school buildings, and university buildings would have been rebuilt with passive thermal design. The hospital, parks, location of roads, city centre, rail, port, and surrounding farmland would still be used. New Zealand's national grid would still be working, and the hydro-power and geothermal generation would be at least 80% of the current level of electricity supply. We also determined that communications through high-speed internet with the rest of the country and the rest of the world would be sustained.

In our exploration of this future BHS Zone, we found that a small-town centre had been created so that residents could cycle to 50% more offices, shops, jobs and markets without having to travel to the next closest centres. We found that people had reclaimed much of the roadway space for some high-density apart-ment housing in the centre, and for other uses like new forms of urban hot-house agriculture to supply fresh vegetables, fruits and eggs to the town market. Annual crops like potatoes and grains were brought to the city by train from the outlying farmland. There were many people working in agriculture, food preservation and management.

The country we found in the future BHS Zone had an extensive electric rail net-work, and all towns had hybrid-purpose trams. Freight was one purpose – moving pallets of goods and wastes during the night and early morning. The trams also moved people, cycles, strollers and hand-carts during the day. There were electric delivery carts, something like the vehicles currently used for golf carts, and there was a nearly unlimited diversity of electric cycles and motorcycles, all manufac-tured in New Zealand.

A surprising discovery was the number of new jobs in urban gardening, in which about 75% of the formerly ornamental verges, lawns and roads had been taken over by intensive urban agriculture. Another surprising discovery was the micro-manufacturing that produced the goods for local consumer needs, some-times with designs from overseas. There was no disposable plastic, no solid waste going to landfill, and no pollution going into the waterways. Rather, all waste materials were reuseable or compostable and handled by the urban gardeners who were also the waste management contractors for the city. There were new jobs in the tram and rail businesses. There were many new jobs in processing and light manufacturing.

In backcasting the life in this future suburb without personal petroleum vehi-cles, we found that the income currently needed to support the import of fuel and vehicles and build and operate the infrastructure for them would be more than sufficient to support electric tram and trains. We also found that the current waste and inefficiency of electricity use would be reduced by transition to the lower consumption in high-standard residences and buildings and would allow growth of electricity for the rail and tram systems. In order to maintain the access to computing and communications, a transition would also be needed in much longer-lasting electronics, and they would be used primarily in production and operations – not in entertainment. Thus, there would be new types of businesses serving the social connection needs, for example, dance halls, roller skating rinks, clubs of many different sorts and performing arts theatres.

One trigger we found for the transition was the high school students' value shift away from cars. The trend was just starting in 2003, but it has continued to grow. Younger people's delay in adopting driving behaviour has been part of the reason for the peak and decline of travel demand (number of kilometres driven per person), which has confounded transport planners in recent years. The other trig-ger we foresaw was peak oil and the price volatility and pressure for change that would occur (Krumdieck et al. 2010). The oil price spike and the global economic

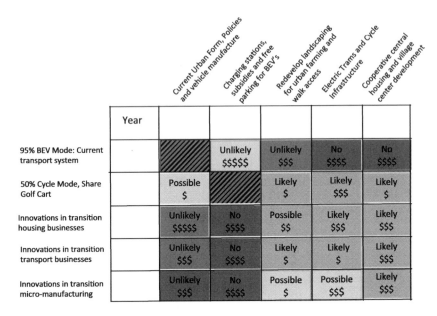

FIGURE E5.3 The possibility space for Christchurch 2100 shows that innovative redevelopment of residential housing systems, electric trams and cycles, and re-purposing of urban landscaping are the best opportunities.

crisis of 2008, as well as increased awareness of global warming's impact on climate change, have likely been contributing to the travel demand decline in most countries, not just New Zealand.

We used the strategic analysis approach to organize all of our shift project concepts and evaluate the opportunity space, as shown in Figure E5.3. One shift project concept that we selected was a pool car service that could allow access to cars when a car was actually needed but would allow someone to shift to a largely car-free lifestyle with all of the financial benefits. The pool car service would be a business that would manage the fleet, maintain the cars, book them in advance, and also deliver them to people when needed. Recall that this brainstorming idea was in 2003 before iPhones, apps and Uber. The opportunity for a pool car service could be realized immediately but the only way to develop it would be to just start it and learn by doing.

Another shift project was well-engineered and ergonomic commuter bikes designed for women, and electric cargo bikes. The opportunity would be to work with a bike manufacturer to develop these new models and learn from test marketing. Another shift project would be a new kind of urban property development that catered to people who wanted access to local markets and products, a walkable environment and a car-free lifestyle. The opportunity for a new urban village development would be harder to realize unless an accessible area of the city could be identified where the property could be available, like a brownfield site of an old factory. The first such development could be part of a transition in city planning and infrastructure design.

We felt that the engineering tools for the kind of urban redevelopment that would be needed were not available, and we embarked on research projects to develop urban accessibility modelling (Rendall et al. 2011). Rather than thinking of a city as a grid of streets and parking, the city could transition into patches of urban villages or town centres interconnected with public transport. The other novel shift project from our brainstorming session was the idea of hybrid design for electric trams, and a type of electric handcart pallet mover that could triple the value proposition for building trams. Trams could carry passengers and move goods from manufacturers, agricultural areas and warehouses to the urban villages and they could remove waste materials back to the manufacturers and areas outside the city. This would mean that the investment for the tram would be better if it connected between the town centres and the industrial and agriculture areas that supported the city, and the tram could collect ticket fees for three trip purposes instead of just one.

This example illustrates how the Transition Engineering methodology leads to discovery of shift projects that are indeed part of the future transition scenario.

6 Economic Decision Support

Borrow from the best, learn from the rest.

This chapter presents a range of approaches for evaluating the costs and benefits associated with energy extraction, production, distribution and use. Conventional economic ideas have developed during the past century of unprecedented population growth, and expansion of manufacturing, building, and consumption. The economics of energy transition introduces a new problem – energy constraint. The price of the energy reflects the cost of supply to some extent, but it is heavily dependent on the nature of the markets and the economy at the time. The change of the value of money with time makes it difficult to use cost of energy as a basis for long-term decisions. Think about the price of petrol 10 years ago. Do you remember purchasing fuel for a car? Does the price you paid at that time affect you now? Think about paying your electricity bill 10 years from now. Is the unit cost 10 years in the future relevant to you today? The answers to these questions are very likely 'no'.

The price of energy is important in terms of the way it affects your access to activities and services *at the time of purchase*. When we purchase and consume energy, the relevance of the price of energy diminishes quickly away from the present, but the importance of access to activities, the emissions generated and the reduction in reserves due to the present consumption actually increase with time. Put another way – the price you pay at the pump today to fuel your car is only important for a few moments when you hand over your credit card. But extraction of the petroleum, construction of a city designed for cars, and the CO_2 emissions you generate using that fuel are increasingly important for hundreds of years.

First, we will set out the standard methods of economic analysis, and then explore some of the problems with the current thinking about values and costs. The implications of conventional economic analysis on sustainability through use of the discount rate will be examined, and we will explore why we are intentionally future-blind. Key concepts include inflation, time value of money, levelized cost of energy (LCOE), lifecycle assessment (LCA) and input/output (I/O) analysis.

6.1 COST OF ENERGY PRODUCTION

All energy is free, but not very useful. All useful energy comes at a price. The price must be higher than the cost incurred in making the free energy useful and available. You can use free wind energy by opening your window for ventilation, or enjoying a tailwind on your bicycle. You can also use an electric fan and an electric bike powered by wind, but it would not be free. A power generation company would need to invest in exploring wind resources, acquiring permits and building a wind farm, and the company would need to sell the power to a utility at a profit. The government would have to support construction of a national grid and regulation of the electricity market. Wind is free; wind-generated electricity is not. It doesn't matter very much what any particular energy conversion and distribution system costs, as long as you pay for it. Trees convert sunlight, water, CO_2 and minerals into wood, for free. Firewood delivered to your door is a useful fuel, and it is not free. Wood pellets are an even more useful form of wood fuel, and they are definitely not free. Government subsidies for renewable energy like wind and wood may reduce the price to particular end-users, but the public investment came from taxes and from debt – and again, was not free.

There is a significant difference between the cost of energy extraction, production and distribution, and the price to end-users. Consumer prices must be high enough to cover all of the costs of the energy supply system, including service of debt, replacing and repairing existing conversion plants, and building new plants. As a transition engineer you will henceforth never refer to any form of fuel or electricity as 'free' or even 'low cost'. You will learn what kinds of development issues affect supply chain costs. You will learn how higher risks in any of the supply chain processes affects the consumer price. You will appreciate that, if the price that the market will bear is not sufficiently greater than the supply chain costs plus the profit margin for a particular energy resource, then that resource will not be developed, even if it is technically feasible. Finally, you will understand that 'free' solar, wind or other renewable energy will never be available without 'cost'.

Electricity is the most useful form of energy, and it will never be free. Electric power plants have been built and operated for over a century. The capital investment costs include engineering, materials, land, and equipment and the cost of building the plant in current dollars. Figure 6.1 shows a broad review of the capital costs of electric power plants in USD/kW generation capacity, based on historical data. The review is updated annually by Lazard, a well-known financial advisory and asset management firm (Lazard 2015).

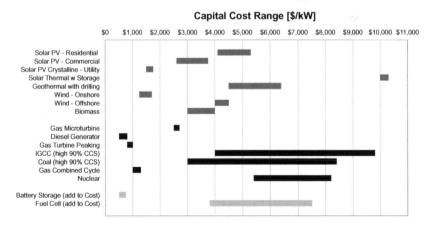

FIGURE 6.1   Range of capital costs in USD per kW generating capacity for electric power generation technologies in 2015 with no subsidies. High-end coal includes carbon capture with no storage. (From Lazard, Ltd., *Energy Technology Assessment*, Lazard, Ltd., New York, 2009, www.lazard.com; Palz, W. and Zibetta, H., *Int. J. Solar Energy*, 10(3–4), 211, 1991. With permission.)

6.1.1 UTILIZATION AND CAPACITY FACTOR ECONOMICS

A power plant cannot operate 100% of the time. All power plants require shutdown for safety checks and scheduled maintenance. A plant may also shut down due to faults, or in the case of wind and solar, when the resource is not available. If the power plant generates power at full capacity only 80% of the time, then the return on investment (ROI) would be lower than if the plant produces power 90% of the time. The utilization factor, U, is calculated by comparing the actual energy production to the potential production for full utilization of the plant:

$$U = Q_e / \left(RC \times 8{,}760 \text{ hr/yr} \right) \tag{6.1}$$

where:
   U = utilization factor (%)
   Q_e = total actual electric energy production (MWh/yr)
   RC = rated electric generation capacity of the plant (MW)

Utilization is an important factor for assessing the economic efficiency of the various assets in a utility or any business. High utilization indicates that the capital investments in capacity are in balance with the requirements. Low utilization indicates unnecessary investment in capacity compared to demand. Renewable power generation like wind and solar have low utilization factors as do gas peaking power plants that may sit idle for many hours of the year when baseload generators can provide the power needed. Figure 6.1 indicates that the capital cost of natural gas combined cycle (NGCC) and wind turbine generation (WTG) can be comparable. The NGCC power plant is a baseload generator with utilization factor around 80%. Utility scale wind farms have utilization factor less than 30%. An investment of $150 million would be needed for 100 MW generation capacity of either wind or NGCC. The unit price that the utility can charge is the same whether the electricity is produced by wind or gas. However, the gas plant generates NGCC_Q_e = (100 MW × 0.8 × 8760 hr/yr) = 700,800 MWh and the wind farm generates WTG_Q_e = (100 MW × 0.3 × 8,760 hr/yr) = 262,800 MWh. The power company will obviously have much less electricity to supply to the market from the wind farm for the same capital investment as an NGCC plant. Of course, the gas plant has added cost of fuel, but you can see how investment decisions in the past have favoured gas, hydro and coal power plants, which have high utilization factors.

The capacity utilization factor (CU) compares the annual generation from the plant to the generation from the plant when it was called upon. The capacity factor is a measure of how reliable the plant is in meeting the market demand:

$$CU = Q_e / \left(CC \times HC \right) \tag{6.2}$$

where:
   CU = capacity utilization factor (%)
   Q_e = total actual electric energy production (MWh/yr)
   CC = called-for capacity (MW)
   HC = hours of called-for capacity (hr/yr)

A peaking gas power plant is designed for coming on line quickly to meet peak loads at any time to maintain voltage and avoid power outages. Peaking plants can be built to provide power for as few as 100 hours per year during the critical peak periods. If a peaking gas plant delivered the capacity called for each hour it was called, then the capacity utilization factor would be 100%. Utilization factor for solar PV is usually under 20%. If you have a company that installs solar PV on retail centres with the objective to meet daytime peak demand for air conditioning in the retail building in the summer, then the capacity utilization factor could be much higher than the utilization factor. Be aware that utilization factor or capacity utilization factor can be defined in different ways for different reasons, but the concept of comparing the investment in capacity to the electricity actually delivered is a useful performance measure.

6.1.2   LEVELIZED COST OF ELECTRICITY PRODUCTION (LCOE)

LCOE is the investment cost, plus maintenance and cost of fuel, per unit of electric energy produced over the lifetime of the plant:

$$\text{LCOE} = \left[\sum C_{cap\_i} + \sum C_{man\_i} + \sum C_{fuel\_i} \right] \Big/ \sum Q_e \qquad (6.3)$$

where:

LCOE = levelized cost of energy ($/MWh)
C_{cap_i} = capital costs in year, i
C_{man_i} = operation and maintenance costs in year, i
C_{fuel_i} = fuel costs in year, i
Q_e = total electricity generation over lifetime

Figure 6.2 shows the ranges of LCOE without subsidies from conventional and alternative energy technologies (Lazard 2015). The cost estimates for nuclear power do not include a facility for disposal of radioactive waste or the decommissioning of the plant. The high LCOE of solar, and peaking gas relative to baseload geothermal, biomass, coal and natural gas is in part due to the low utilization factor. The LCOE from a fuel cell is not from industry data as there are no such facilities. It assumes that hydrogen is generated from natural gas and would be approximately three times higher if the hydrogen was produced by electrolysis from water. The integrated gasification combined cycle (IGCC) is only based on demonstration plants and is assumed to include the cost of carbon capture but not compression, transport and storage. The cost of battery storage would need to be *added* to the generation cost.

Extraction companies access vast fossil resources by paying for mineral rights without having to purchase the land, and local governments collect mineral royalties on production. The LCOE is normally a simple benchmark and does not include upstream costs for energy prospecting and mining, or downstream costs for disposal reclamation or pollution mitigation, and these costs are sometimes reduced by subsidies. Refining and generation costs are included, but LCOE calculation normally neglects financing charges on capital.

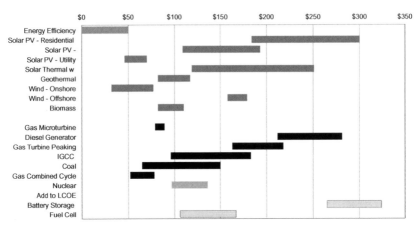

FIGURE 6.2 LCOE comparison showing high discounted levelized cost for peak demand plants like diesel generators, gas turbine peaking plants and utility battery storage. (From Lazard, Ltd., *Energy Technology Assessment*, Lazard, Ltd., New York, 2009, www.lazard. com; Palz, W. and Zibetta, H., *Int. J. Solar Energy*, 10, 211, 1991.)

6.1.3 Payback Period

The payback period (PP) is the time required for the cumulative revenues (or savings) from an energy conversion system to equal the initial costs. PP is determined by:

$$PP = \frac{C_o}{B_n} \tag{6.4}$$

where:
PP = payback period (years)
C_o = capital and installation cost ($)
B_n = benefit in a year, n, e.g. value of energy saved per year ($/year)

This is also called the simple payback period because it assumes that the benefits are the same in each year, and it does not include financing costs or the time value of money, also known as the discount rate. This approach is acceptable for preliminary estimates if PP is short, say, less than 4 years. For a more precise estimate, the time value of money, the inflation rate, and the escalation in fuel costs or maintenance costs can be considered.

6.1.4 Risk: The Cost of Time and Uncertainty

Developing an energy conversion plant requires specialist engineering and materials, insurance, regulatory consent, and safety certification, as well as land and water. Failure to secure any one of these requirements can derail the whole project,

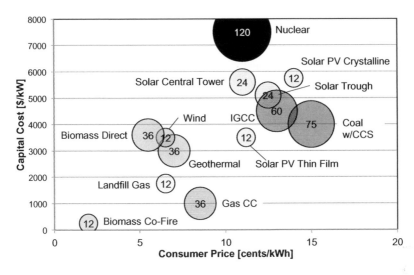

FIGURE 6.3 Consumer cost, capital cost, and construction times for various supply-side alternatives (circle size indicates construction time in months). (From Lazard, Ltd., *Energy Technology Assessment*, Lazard, Ltd., New York, pp. 11–13, 2009, www.lazard.com. With permission.)

and getting them takes time. Figure 6.3 gives estimates of the development time indicated as the size of the circle in number of months. Construction time is an important factor to investors as it represents development risks (Lazard 2009). The time to design and install solar PV systems is less than 12 months, if the system is for summer demand peak reduction and placement is on the roof of a commercial building that can use the power directly, with no feed-in to the grid, no battery storage system, and no need for architectural approval. Landfill gas generators also have a relatively short timeframe if the resource has previously been measured and characterized. The regulatory processes to develop landfill gas are short because of the previous industrial use of the site, and landfills are usually in good proximity to grid tie-in.

The level of complexity involved in integrating the new generation into the landscape and the grid can cause longer construction times. Wind farms can be constructed in under a year, but there are often issues with acceptance of the site, being able to install the needed firming supply in the utility, and grid tie-in if the wind farm is sited in a remote location. The longest construction time is for, of course, nuclear power plants. The 120 months in the figure would be for a plant that has received approval and has government guarantees of financing and limits on liability in case of an accident. After the Three Mile Island nuclear power plant accident in 1979 in the United States, orders for new power plants were cancelled and there were no new projects started for 3 decades. The V.C. Summer project was proposed in 2007 for two large nuclear reactors in South Carolina, United States. The engineering and design process became complicated, to a certain extent by the lack of nuclear engineering capability. In July 2017, the state-owned utility made the decision to abandon the project, which was less than 40% completed and had already cost \$9 billion. The utility had raised consumer electricity prices several times to cover the cost overruns of the project. The project contributed substantially to the bankruptcy of the reactor designer and supplier, Westinghouse, and now accounts for 18% of the electricity bill for residential customers. This example illustrates what is meant by 'risk' in the energy conversion sector. In the South Carolina nuclear power plant case, the risk was borne mostly by residents of the state because it was a public utility.

The energy development sector is also sensitive to the risk posed by public opposition to projects. Geothermal energy taps into naturally heated groundwater, but in many countries, the natural hot springs are sacred sites. In the case of one of the largest geothermal resources in the United States, the thermal resource is protected within the Yellowstone National Park. Wind turbines do not cause permanent large-scale damage, but wind farms face stiff opposition from local landowners due to the way they affect the visual quality of the natural landscape. Of course, coal mining, extracting tar sands, drilling for oil in the deep ocean, and hydraulic fracturing have well-understood risks of environmental pollution. These extraction operations, such as the Keystone oil pipeline project to transport tar sand oil from Alberta, Canada, to refineries in Texas, United States, often face development delays from protesters and political reluctance.

Insurance and liability are major issues with high-risk power plants like offshore energy conversion and nuclear power. The time-to-build estimates for IGCC and coal with carbon capture and sequestration (CCS) are large according to Lazard, who cautions that no commercial facilities have been constructed and there are no suppliers for the equipment, so the project development time may actually be prohibitively long for many years to come. Note that hydrogen fuel cells are not energy production technologies, so the costs of fuel cells, electrolysis, compressors and storage tanks would be *added* to the cost of producing or using the electricity if hydrogen were used. Batteries are also an additional cost to any system.

6.2   ENVIRONMENTAL COSTS

6.2.1   EMISSIONS FROM FOSSIL FUELS

There are constraining forces on energy conversion associated with environmental impacts. The energy provided to the end-user has a market price, which may or may not include waste disposal or emissions costs. Environmental costs will eventually be paid by society through government spending to try to fix the problems. The costs of climate change are increasing with severity of storms and losses due to drought and heatwaves. The expansion of the range of tropical diseases is another example of social costs due to energy use.

Environmental costs are often referred to as external costs or externalities. Conventional economics is based on the idea that price is the mechanism by which supply and demand are balanced. Economists argue that putting a price or a tax on externalities would stimulate the technology change to reduce the environmental costs. Only political decisions could levy and enforce effective costs on externalities, but this has not happened very often. In the case of greenhouse gas (GHG) emissions, some cost instruments have been introduced, but the carbon emissions pathway, as business as usual (BAU), has not objectively been altered. Using the flipped perspective, the work of energy Transition Engineering is to change technologies and operations to step down GHG emissions and then find the financial justification for the changes internally.

The energy sector is directly responsible for 70% of the GHG emissions that cause global climate change. Carbon dioxide (CO_2) has been known to absorb heat and act as a GHG for more than 100 years. The current level of atmospheric CO_2 above 350 parts per million (ppm) is known to have shifted long-standing climate patterns in most areas of the globe (NASA 2015). The scientists who observe changes in everything, from polar ice to growing seasons, to seawater chemistry, are calling for immediate and drastic reduction in CO_2 emissions so that climate systems do not become catastrophically destabilized.

The Intergovernmental Panel on Climate Change (IPCC), in its Fifth Assessment Report (2014), outlines the likely risks and changes if the production of fossil fuels and the resulting GHGs is not curtailed. The IPCC has set a value of 800 Gt cumulative fossil carbon as the climate failure limit. This is not a target. To date, more than 550 Gt of fossil carbon has been extracted and burned. The current fossil fuel production rate is 11 Gt-C (40 Gt-CO_2) carbon per year. The IPCC is calling for an 80% reduction in GHG emissions by 2040, with drawdown of atmospheric carbon dioxide via negative net emissions by 2050 to keep the global temperature increase below the potentially catastrophic level.

In 2014, the International Energy Agency (IEA) proposed four actions that would be economically beneficial, decouple economic growth from energy consumption and avoid exceeding global warming above 2°C:

- Eliminate the $550 billion/year government subsidies for fossil fuel production in all countries.
- Focus on energy efficiency to stop and reverse demand growth.
- Increase investments in low carbon shifts by a factor of four.
- Increase CO_2 prices and government support for CCS.

The lock-in effect for GHG emissions refers to the infrastructure investment that has already been made in using fossil fuel. The obvious locked-in infrastructure investments are coal power plants, highways, and airports. But the urban form, especially the geospatial distance between residences and destinations, also requires use of the car. The IEA is calling for investment that reduces carbon emissions. It states that many governments perceive that they are saving money now by not investing in efficiency, demand reduction or renewable energy. However, the IEA analysis shows that every $1.00 not invested now to keep CO_2 atmospheric concentrations below 500 ppm will require $4.30 of investment in the future to deal with the costs of climate change. Governments around the world are politically stalled on imposing the sufficient carbon price of $140/t-$CO_2$. One ton of fossil carbon burned with oxygen produces 3.67 tons of CO_2, and subbituminous coal is 72%–76% carbon by weight. Thus, the IEA is estimating that an additional externality price of ($140/t-$CO_2$ × 3.76 t-CO_2/t-C × 0.74 t-C/t-coal) = $390 per ton of coal is warranted to achieve the necessary shift away from coal. For reference, the price of subbituminous coal in the United States over the past five years has ranged from $65 to $40 per ton. This effective price of externalities would result in a ten-fold increase in the price of coal-generated electricity, which would definitely result in the closure of coal-fired power plants.

A recent trend amongst some large companies is to impose an internal CO_2 price on their own fuel and electricity purchases. They then set up a fund that employees bid into to pay for projects that reduce the company's electricity, carbon-intensive materials, air travel or fuel consumption. In this way, the net effect of the change on profit is zero, but the company has found a way to finance the transition to the much more competitive position of using less energy and materials per unit of production.

6.2.2 HEALTH COSTS FROM ENERGY CONVERSION

Soot, ash and SO_2 from coal, wood and diesel, and NO_x from petroleum and natural gas combustion are the major pollutants that directly cause negative health effects. Combustion science and air pollution engineering have provided great improvements in air quality since the 1970s in countries that have instituted clean air regulations. Acid precipitation from the sulphur dioxide and the accumulation of mercury from coal combustion are an ongoing problem in aquatic ecosystems. Mercury is a potent neurotoxin. Today, it is estimated that one in six American women of childbearing age have blood mercury levels that exceed the safe exposure limit for foetal development. This means that as many as 600,000 babies are born each year at serious risk of prenatal methylmercury exposure, which has been determined to cause permanent and irreversible impairment with language, memory, attention, visual skills, learning and reasoning (Mahaffey et al. 2004).

Every state in the United States has issued health advisories banning fishing and consumption of fish due to toxic mercury exposure in at least one lake or river. The largest source of mercury pollution in the United States is coal-fired power plants, but combustion of coal in industry has also caused some very high local mercury contamination problems (US EPA 2005). Mercury accumulates in the fatty tissues and oils in all animals and is never broken down or dispersed in the environment. Thus, release of mercury into the atmosphere from coal combustion is a serious sustainability issue for humans, agriculture and wild animals, particularly aquatic predator species like tuna, seals and dolphins. Coal-fired power plants in the United States are currently not required to limit their mercury emissions, even though the abatement technology is available. The US Environmental Protection Agency (EPA) (USEPA 1990) set new standards on mercury emissions under the Clean Air Act. However, the Obama era Mercury and Air Toxics Standards (USEPA 2011) are now under review due to recent political change. As with nuclear energy and climate change, these types of long-term cumulative emissions are difficult to deal with in economic terms and usually require social pressure for regulation and change in practices by the emitters.

6.3 CONVENTIONAL FINANCIAL ANALYSIS

All projects require cost estimating and financial analysis, so it is necessary that you learn about money and develop effective strategies for working with investors to gain financing. Our concept of money is based on the idea that money is a real and physical thing that can be exchanged for goods or labour. Gold and silver have had value since prehistory, most likely for the same reason they do today. You can't make your own, so there is a limited supply of both metals, and they are durable: they won't rust, burn or break. The first coins were made more than 2,500 years ago. Typically, a government's wealth was held in a treasury, and representations of that wealth in bronze and other metals served as proxies for that wealth. Coins, paper bills and now electronic digits are called fiat money, meaning that the money has no intrinsic value in itself. Rather, the value is declared by a government, which relies on the belief in that value by the people who use the currency.

After the US Civil War, the treasury printed $354 million in paper money that was not redeemable for silver or gold, but that was worth the amount of gold held in the Federal Reserve. This $354 million was capped and remained the amount of US dollars in circulation for nearly a century. After World War II, the nations of the world made an agreement to use the US dollar as the standard for international exchange. This was largely because the US national treasury held most of the world's gold, so the value of the dollar was actually the most stable.

The global money system changed radically in 1971 when the US dollar standard was disengaged from the quantity of gold held by the US government. Since that time, banks could lend money into existence rather than lending out money deposited by others, as the older savings and loan institutions were required to do. When debt is created, faith in the future payment of interest on the debt is the primary basis for the money supply. Since the 1970s, the amount of US cash (paper money) in circulation has grown to more than $1.2 trillion, the majority of which is actually held by people outside the United States. This system was severely tested in 2007 and 2008 when government debt, secured by future tax revenues, was required to back the debt created by banks. Since 1971, the money supply has grown exponentially, mostly through debt owed on bonds issued by the government. In addition, consumer price inflation was higher than any previous time except during wars. In 1971 the US gross domestic product (GDP), or all of the money spent in the country, was roughly 34% of the national debt (mostly in government bonds). In 2015, the US GDP had grown to $18 trillion, but the debt had soared to 99% of GDP. This new experiment with the monetary system has coincided with deregulation of the banking industry. Since the 1980s, there has also been deregulation of utilities and shifting of the objectives of the power supply system operation from providing affordable, reliable electricity to all end-users to generating revenue for shareholders.

Investments in energy systems depend on the beliefs, perspectives, and expectations for returns of the investor. In the later sections, we will explore how innovations in financial and economic perspectives offer opportunities for energy transition investments.

6.3.1 TIME VALUE OF MONEY

The basic idea of the time value of money derives from the way capital (money) is generated. One way to get capital to purchase equipment or invest in energy transition is to borrow money from a bank. If you borrow money, you must pay it back over time, plus interest. If you deposit money in a bank, it earns interest over time. If you borrow money, then your available revenues that could earn interest are reduced. Money to build a power plant is normally too large and the payback too long to be secured by bank loans. The funding for infrastructure can be raised through selling of bonds to investors who expect to get paid back plus a share of returns. Investors expect that the new investment would be used to grow the company and provide higher profits. Investors in bonds expect that the tax revenues of a growing country will also provide returns. Any way you look at it, the capitalist system requires growth in order to finance the capital expenditures.

A company can accumulate and hold profit over time and then invest this capital into new capacity or replacement of old equipment that has depreciated over time. This capital investment, called CAPEX, then produces benefits and has an internal rate of return (IRR) in profitability of the company. In the annual budget, the depreciation amount for each capital item is counted as a cost of doing business. This internal depreciation charge is how the CAPEX expenditures are financed. Let's say that a pipe manufacturing company must have a hardness tester. In 1980, the company purchased a tester for $60,000, and the expected lifetime was 20 years. Thus, the depreciation was $60,000/20 = $3,000 per year. In 2000, the company must replace the old hardness tester, but the cost is now $70,000, and for the next 20 years the depreciation will be $70,000/20 = $3,500 per year. Think of depreciation like saving for a car. You could put some of your income in an account each month until you had enough for a car. But then you would need to keep putting money in your account over the lifetime of the car in order to have enough to purchase the next car. If you expect that the price of cars might be higher in the future, then you would want to increase the amount you save. Thus, you need to increase your income in order to keep up with your own progress.

Inflation is another factor that affects the value of money with time. People have used coins for more than 2,500 years, and inflation has occurred when a ruler reduced the precious metal content of coins and issued more coins with the lower value. The word *inflation* was first used in the early 1860s, during the US Civil War, when the US government changed from the gold dollar to the paper dollar, called a greenback. This new greenback started to lose value after it was printed because people had less confidence in the paper than in the gold and silver coins. If people devalue the money you are holding, then they want you to pay more for goods, thus causing prices to increase. Hyperinflation is a recent phenomenon that began when the value of fiat paper currencies issued by a government were not directly related to an amount of precious metal held by that government (Bernholz 2003). The most famous example of the failure of a currency is the Weimar Republic hyperinflation. Germany suspended the gold standard for the mark and printed more paper money to pay for its expenses in fighting World War I. The reasoning was that, after winning the war, the Weimar government could take gold from the losers and pay off the debts to armament makers. However, the victorious Allies imposed massive reparation payments on Germany in gold, and the government printed large amounts of money to pay debts and civil workers. This led to devaluation of Germany's paper mark, which in turn caused hyperinflation of the cost of goods. The German currency went from trading at 90 marks per US dollar before the war to 4.2 trillion marks per US dollar in 1923. The hardship that the German people endured because of this hyperinflation should be understood by anyone who currently believes in the value of money.

Today, we think about inflation as the increase of prices across the whole economy, but there are four key drivers of inflation: finance, food, fuel and speculation.

We have already discussed why the use of debt causes the need to increase revenues, but increasing prices is a more profitable way to increase revenues than increasing capacity. If you expand your production, then you have to finance even more capital. But if you charge more per unit of production, you increase revenues without adding to the costs of doing business. The price of electricity has increased more with increased capacity. Each investment in new capacity has been, and will be, more costly and higher risk than past investments. Thus, the financing on new generation and transmission capacity will demand higher interest, and consumers will cut back on demand, so the price must escalate ever faster.

The other source of inflation is scarcity. Food and fuel are essential goods, so people and organizations are willing to pay more if they perceive scarcity. In the current global trade environment, this means that countries that can afford to pay more can cause scarcity and high prices in other countries. Many urban dwellers around the world are facing historically high housing prices, and the inflation in property price is a result of speculation and perceived scarcity. Objectively, a house built in 1992 on a 500 m^2 urban lot has not changed in real value in the last 30 years. In fact, the house would need maintenance and updating in order to maintain the real value as a residence. Appreciation is the higher price that someone is willing to pay for your house in the future. The belief that the house price will continue to be higher in the future leads to willingness to pay more for the same house each year. This belief in future returns is called speculation. In 1992, the average value of a house in the United Kingdom was £50,000; in 2007, the average price was £180,000; and in 2017, the price had escalated to £219,000 (Halifax House Price Index, 2017).

6.3.2   CASH FLOW: COSTS AND BENEFITS

The cost and the benefits of a project can be expressed either in current dollars or constant dollars. Actual cash flows observed at a given time are called current dollar cash flows. They are the actual number of dollars in the year that the cost or the benefit is incurred. A cash flow in a given year is equal to the benefits minus the costs:

$$F_n = B_n - C_n \qquad (6.5)$$

where:
   F_n = the cash flow in the year n in the current dollars of year n
   B_n = benefits: the revenues or savings in year n
   C_n = costs: the capital expenditures, expenses or losses in year n

Current dollar cash flows change over time because of inflation (or deflation). Inflation-adjusted dollar cash flows are found by using inflation as a growth rate:

$$\bar{F}_n = F_n (1 + j)^n \qquad (6.6)$$

where:
   $\bar{F}_n$ = inflation-adjusted value of the cash flow in future year n
   j = annual inflation rate

Equation (6.6) assumes a constant rate of inflation. Past inflation has been highly volatile, but it is accurate enough when looking at future values to use the historical average, e.g. 2%, as a base case, and investigate scenarios for higher rates as part of the risk analysis. Increasing pressures on resources, insecurity in oil supply, carbon taxes, and exponential growth in the money supply present risks for considerably higher inflation for energy prices in the future. Sustainable energy analysis should explore the effects on businesses with constant or growing energy demand in the case of future inflation rate scenarios with 2% per year average, 8% per year high and 50% in one year representing a fuel crisis. Engineering calculations are used to determine the size and performance of the energy system investment in future years.

6.3.3 NET PRESENT VALUE (NPV) AND DISCOUNT RATE

NPV is used to compare options with different lifetimes and different costs and benefits in different future years. The basic idea is that NPV brings the cash flows in future years to the present through use of a discount rate determined from the investor's internal time value of money. The calculation process uses a spreadsheet to place revenues and costs in future years, applies Equation (6.7) to each future cash flow year, and sums the discounted present values:

$$NPV = \sum_{n=0}^{N} \frac{F_n}{(1+d)^n} \tag{6.7}$$

where:
NPV = net present value of the cash flows in future years
F_n = cash flow received in year n
d = discount rate
N = analysis period in years

The NPV of a cash flow in just one future year is NPV_n. The present value of a benefit in a future year is denoted $PV_n = B_n/(1 + d)^n$ (the equation for costs is similar). The discount rate can be thought of as the return that a company normally expects when it invests capital in expanding or improving its business. The discount rate used by public-sector projects is typically lower than for large companies. High discount rates are expected for higher risk investments. The discount rate is determined by the organization that is making the investment decision and is thus provided to the engineer for carrying out the analysis.

Understanding how the NPV discounts the future is essential for sustainability analysis. For example, if we are considering an investment in retrofitting a building to reduce energy use, then we calculate the reduced energy demand and the cost savings before and after the retrofit as the benefits in future years. If the discount rate is given as 10%, and we calculate energy savings of $10,000 per year for an office building, the present value of the savings in year 15 is $PV_{15} = \$10,000/(1 + 0.1)^{15} = \$2,393.92$. Thus, this perspective on the time value of money causes the future savings to be devalued or discounted. Note that the analysis also devalues the future costs, which means that overall, not investing in change tends to come out as the best option. You would be correct in thinking that this sounds like what has been happening since the 1980s.

Figure 6.4 shows the current dollar cash flow diagram that is often used to illustrate the performance of the energy sector investment. It is clear to see why an Excel spreadsheet is used for this kind of financial analysis. In Figure 6.4, a 4-MW commercial-scale PV array is to be installed on the roof of a very large factory in Reno, Nevada, United States. The capital cost of the array is $3,000 per kWh, and the time to build is 12 months. In the first year, the system generates 6,752,441 kWh of electricity, with a capacity factor of 19.3%. The electricity produced is used in the factory to reduce power purchased from the grid at $0.097/kWh, with annual inflation assumed to be 2%. There is an inverter replacement and maintenance needed at 7 years, which is 15% of the capital cost, and disposal cost at the end of 15 years is

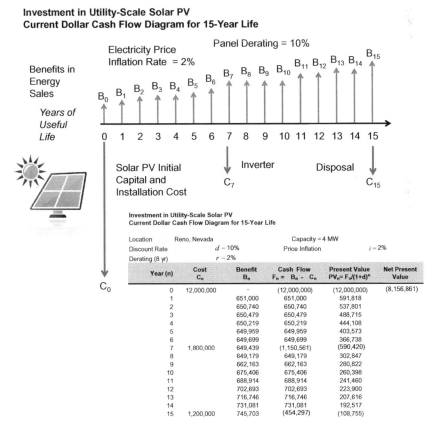

Investment in Utility-Scale Solar PV
Current Dollar Cash Flow Diagram for 15-Year Life

Location	Reno, Nevada		Capacity = 4 MW	
Discount Rate		$d = 10\%$	Price Inflation	$i = 2\%$
Derating (8 yr)		$r = 2\%$		

Year (n)	Cost C_n	Benefit B_n	Cash Flow $F_n = B_n - C_n$	Present Value $PV_n = F_n/(1+d)^n$	Net Present Value
0	12,000,000	-	(12,000,000)	(12,000,000)	(8,156,861)
1		651,000	651,000	591,818	
2		650,740	650,740	537,801	
3		650,479	650,479	488,715	
4		650,219	650,219	444,108	
5		649,959	649,959	403,573	
6		649,699	649,699	366,738	
7	1,800,000	649,439	(1,150,561)	(590,420)	
8		649,179	649,179	302,847	
9		662,163	662,163	280,822	
10		675,406	675,406	260,398	
11		688,914	688,914	241,460	
12		702,693	702,693	223,900	
13		716,746	716,746	207,616	
14		731,081	731,081	192,517	
15	1,200,000	745,703	(454,297)	(108,755)	

FIGURE 6.4 Graphical representation of a current dollar cash flow series for investment in a 4-MW utility-scale solar PV array in Reno, Las Vegas, Nevada and the Excel spreadsheet used for financial analysis.

10% of capital cost. The solar panels have declining efficiency of about 2% per year until a total derating of 13%. The NPV is minus \$8.1 million with a discount rate of 10%. This means that the net present value is actually a cost.

6.3.4 INTERNAL RATE OF RETURN (IRR)

Investors compare potential investments by the rate of return they generate, and companies compare different uses of CAPEX by the length of time they take to pay back. The IRR analysis is commonly used for accept/reject decisions by comparing the IRR for the proposed project with a minimal acceptable hurdle rate. The IRR of an investment that has a series of future cash flows (F_0, F_1, ..., F_n) is equal to the rate of ROI that sets the NPV of cash flows equal to zero. IRR implicitly assumes reinvestments of any return at the IRR rate and is therefore not recommended for ranking projects. However, it has the advantage of giving a designer a quick comparison of return for assessment of energy management project options as either achieving the hurdle rate or not:

$$\text{NPV} = 0 = \sum_{n=0}^{N} \frac{F_n}{\left(1 + IRR\right)^n} \tag{6.8}$$

6.3.5 PROCESS CHAIN ANALYSIS (PCA)

PCA accounts for the energy and resources needed and the waste produced over the different production steps. Figure 6.5 shows a basic schematic of the PCA methodology for production of sheet aluminium (Boercker 1978). Many companies are now examining the supply chains for the goods they use and looking to select suppliers with a lower carbon footprint, lower environmental impact, or better worker relations in the process chain. Thus, PCAs are becoming a common requirement for sustainability engineering. A survey of supply chain professionals in 2017 found that 97% of organizations place a high level of importance on sustainable procurement, with results of driving revenue and mitigating business risk (EcoVadis 2017). Nearly all companies are looking for new innovative ways to track and analyze the energy intensity, efficiency and emissions of their suppliers and for ways to communicate and instigate emission reduction shifts in their suppliers.

6.3.6 LIFECYCLE ASSESSMENT (LCA)

LCA is a tool to assess the energy, water and material use and emissions for the production, use and disposal of a product. LCA is also called cradle-to-grave analysis, and it is used to compare products with the intention of informing policy decision making. LCAs for many materials and end-use appliances are available, in particular from Argonne National Lab and the National Renewable Energy Lab (NREL) in the United States. The Greenhouse Gases, Regulated Emissions and Energy Use in Transportation (GREET) database is free for download (www.greet.es.anl.gov).

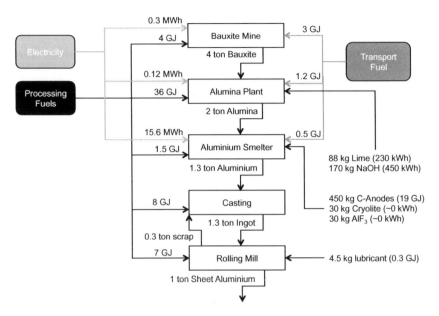

FIGURE 6.5 Quantitative PCA of the energy used in the production of primary aluminium. (From Boercker, S. W., *Energy Use in the Production of Primary Aluminum*, ORAU/ IEA-78-14, Oak Ridge Associated Universities, Oak Ridge, TN, 1978.)

LCA cannot be used to determine if a product is sustainable in an absolute sense, but it can be used to point out what steps in a production process or supply chain are responsible for the impacts. For example, growing algae as an energy source was a popular idea in the early 2000s. Biodiesel has been produced from algae in research-scale processes using algae ponds with paddle collection, centrifuging, natural gas drying, chemical extraction of oils with hexane, then transesterification with MeOH. Every 1,000 MJ of biodiesel also produces 34 kg of waste products, consumes 20.3 m^3 of water, and requires 6,670 MJ of input processing energy from electricity and natural gas (Sander 2010). This result led the researchers to conclude that more research was needed to find new processes. However, a transition engineer should conclude from the result that producing biodiesel from algae is not sustainable, has too low of an energy return on energy invested (EROI), and therefore is highly unlikely to be part of the energy transition.

The LCA is used to compare possible energy alternatives to the current conventional systems. The two main areas of interest are electricity generation and transportation. Dozens of researchers have studied the LCA of energy conversion technologies. Each study has different boundaries and assumptions and calculation methods. A good resource for getting the big picture and not getting stuck in details is the Harmonization Project from NREL. Harmonization is a meta-analysis approach where all related published work is taken together to form a landscape of possibility as agreed by experts in the field. The lifecycle CO_2 emissions results for different electricity generation technologies show that while solar PV, solar concentrating thermal and wind generation are not zero carbon energy sources, they are all in the range of ten times lower lifecycle emissions than the cleanest coal-fired power generation (Heath and Sandor 2013).

The comparison of energy-consuming technologies is also being examined with LCA meta-analysis. There are no published LCA studies of personal vehicles that show that electric vehicles (EVs) are carbon free options, and in fact there are some models of EVs that have the same lifecycle CO_2 emissions per km as high-efficiency diesel vehicles. But a meta-analysis by Motti et al. (2016) gives the clear picture that the fleet of EVs has the lowest gCO_2/km, about one-third of the average internal combustion engine (ICE) fleet. One flaw of LCA is that there is no way to account for a reduction in driving distance of an existing ICE vehicle as a low carbon policy or investment choice. Driving existing ICE vehicles less is a far superior emissions reduction strategy, given the substantial carbon footprint of an EV.

Total lifecycle cost (TLCC) is a way to translate all of the inputs into monetary terms. The analysis should consider all significant dollar costs over the life of a project, including exploration and mining and end-of-life disposal. These costs are then discounted to a present value using the equation:

$$\text{TLCC} = \sum_{n=1}^{N} \frac{\sum C_n}{\left(1+d\right)^n} \tag{6.9}$$

where:

ΣC_n = all direct and related costs in year n, including financing, operation and management (O&M), and energy costs

d = discount rate

Note that the TLCC in Equation (6.9) uses the discounted future costs. It is possible to use a discount rate of $d = 0$, which is referred to as the simple TLCC; it is simply the sum of all costs regardless of what year they occur.

6.3.7 EMBODIED ENERGY

Embodied energy is a quantitative measure of the energy used in the processing chain for a material or product. Embodied energy for a product is usually estimated from the TLCC for the materials used in manufacture. The equipment and energy purchases of particular sectors are compared to the production of the sector. The embodied energy (kWh/$ product) is the ratio of all energy and embodied energy in equipment to the dollar value of the industry production.

The energy intensity factors in Table 6.1 were calculated for an energy-based input–output model of the US economy (Ballard et al. 1978, Hannon et al. 1985), and adjusted to 2008 dollars. The ratio of the 2008 Consumer Price Index (CPI) to the 1977 CPI is 3.52 (Bureau of Labor Statistics 1998). The CPI gives inflation for consumer prices. The data for 2008 show much lower energy per unit production

TABLE 6.1

Embodied Energy in Selected Materials and Processes, in 2008 Inflation-Adjusted USD

Commodity	1977 kWh/$	2008 kWh/$	Commodity	1977 kWh/$	2008 kWh/$
Primary aluminium	46.4	15.1	Fertilizers	52.7	17.2
Aluminium Casting	20.5	6.7	Glass products	17.1	5.6
Copper Mining	25.4	8.3	Glass containers	23.6	7.7
Primary copper	33.1	10.8	Plastics	36.9	12.0
Copper Rolling	20.7	6.7	Cement	63.5	20.7
Electronic equipment	11.2	3.6	Concrete Product	14.9	4.9
Fabricated metal	13.8	4.5	Lime	69.1	22.5
Iron Ore mining	27.2	8.8	Bricks	37.5	12.2
Steel forging	22.0	7.2	Gypsum products	29.1	9.5
Structural steel	14.5	4.7	Chemical products	27.9	9.1
Battery	20.2	6.6	Carbon products	31.9	10.4
Dams	13.0	4.2	Meat products	12.4	4.1
Generators	9.8	3.2	Textile goods	19.3	6.3
Power transportation equipment	9.3	3.0	Stone products	11.6	3.8
Heavy machinery	10.1	3.3	Wood products	15.2	5.0
Aircraft	6.2	2.0	Water, sanitation	26.4	8.6
Local transport	9.6	3.1	Trucks	10.0	3.3
Highways	17.4	5.7	Bus service	2.9	0.95
Air transport	23.7	7.7	Railroads	10.3	3.4
Pavement	53.3	17.4	Pipeline transport	19.7	6.4

value than 1977. There have been some energy efficiency improvements in many industries, but nearly all of the difference is actually due to money in 2008 being worth less than 30 years earlier. This is a good illustration of the destructive effects of the inflationary economic system. All producers, whether producing dairy products or ingots of aluminium, must continuously expand their production, and thus their energy and resource consumption, in order to even maintain a particular level of income in real terms.

6.3.8 INPUT/OUTPUT (I/O) ANALYSIS

Industries use goods and materials to manufacture their products, which in turn are used by others to make *their* products. In the 1950s, the economist Wassily Leontief presented I/O analysis as a way to understand the complex interactions of production sectors. For energy transition, I/O analysis can help us understand the impact in energy in terms of economic productivity. A table is constructed to catalogue the interdependence of sectors and to illustrate the blow-up result of the causal loops between sectors. The I/O problem is formulated by first setting the target production of three sectors to be supplied to the economy in either energy equivalent units or dollar amounts. Next the inter-sector consumption relations are determined, and the total production from each sector required to satisfy the demand is then given by:

$$\bar{x} = \bar{A}\bar{x} + \bar{D} \text{ or } \bar{x} = \left(\bar{I} - \bar{A}\right)^{-1}\bar{D}$$ (6.10)

where:

x_1 is the total production from sector 1, x_2 is the total production in sector 2, and so on
A_{ij} is the amount of sector i output used to produce a unit of sector j product
D_i is the consumer demand for the production from sector i

Figure 6.6 gives an example of the output and inter-sector consumption. There are three sectors: sector 1 = food production, sector 2 = manufactured goods, sector 3 = energy. The consumer demand is 20 units of food, 15 units of manufactured goods and 18 units of energy. The diagram shows the inter-sector demands. The food sector does not require any food to produce a unit of food, so $A_{11} = 0$. The food sector requires 0.5 units of manufactured goods per unit production, so $A_{12} = 0.5$. The energy sector is consuming 0.3 units of food for each unit of energy produced, so $A_{13} = 0.3$. The energy sector requires 0.3 units of energy per unit of energy production, so $A_{31} = 0.3$. The sectors have to produce enough to satisfy consumer demand plus their own internal use, plus the use by other sectors.

The I/O matrix for a system of the three sectors is shown in Figure 6.6. The demand per unit production between sectors is shown with arrows, labelled with A_{ij}, and

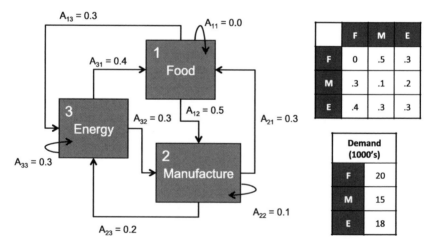

FIGURE 6.6 Example of three production sectors that meet different consumer demands and have inter-sector demand as shown.

organized into an input–output table, which is also the inter-sector demand matrix in the figure and in the linear algebra form below:

$$
\begin{bmatrix} x_1 \\ x_2 \\ x_3 \end{bmatrix} = \left(\begin{bmatrix} 1 & & \\ & 1 & \\ & & 1 \end{bmatrix} - \begin{bmatrix} 0.0 & 0.5 & 0.3 \\ 0.3 & 0.1 & 0.2 \\ 0.4 & 0.3 & 0.3 \end{bmatrix} \right)^{-1} \begin{bmatrix} 20 \\ 15 \\ 18 \end{bmatrix}
$$

$$
\begin{bmatrix} x_1 \\ x_2 \\ x_3 \end{bmatrix} = \begin{bmatrix} 85.0 \\ 68.0 \\ 103.5 \end{bmatrix}
$$

The solution for the total production required in each sector can be found by using the matrix solution in Matlab or other numerical methods. The solution of Equation (6.10) shows the total production required from each sector to meet the demand. The food sector produced 20 units to meet demand and did not have any internal demand ($A_{11} = 0$). The energy sector met a demand of 18 units, but the internal demand for the energy sector was one unit for every 60 units produced ($A_{33} = 0.3$). The effect of consumption between sectors and sector self-consumption on the total production can be quite large. As can be seen from the solution of the I/O matrix, in this example, the energy sector (x_3) would need to actually produce 103.5 units in order for the whole economy to work and to meet an end-use demand of only 18 units.

I/O analysis can be used to explore the effect of different policies and technologies that change total consumer demand and improve efficiency in sector production. This example has random inter-sector demands, but it illustrates why mechanization or efficiency improvements in one sector affect other sectors and the total extraction. The other important conclusion we can gain from I/O analysis is why the reduction of demand for materials and manufactured goods causes reduction in total energy demand, leading to transition from energy demand growth to decline.

6.4   DISCUSSION

This chapter presented the common models used in economic analysis of projects. The main thing the reader should notice is that economics is not an applied science like engineering. Economics is a social and political understanding of production, distribution and consumption of goods and the roles of prices and policies in ensuring that production and distribution are sufficient to meet consumption and to generate wealth. Before the industrial revolution, the main issues were managing the use of common land and water, and fairness in the marketplace. Thomas Aquinas (1225–1274 AD) was a philosopher, but he wrote some arguments about the responsibility for businesses to establish just prices. In the seventeenth and eighteenth centuries, economic thought reflected the dominance of agriculture as the source of wealth, and ownership of productive land by the aristocracy. In this view of economics, the labour of workers is related to the value of things workers produce. The power of trade guilds, and later labour unions, has also influenced the availability, prices and quality of goods. Taxes, borrowing money and providing infrastructure are historical questions of policy that have been recognized to have an impact on production and consumption.

The main purpose of economics was control of the local markets until oil use allowed access to global goods and workers. The field that we recognize as political economics developed after the industrial revolution. The American Revolution and the French Revolution coincided with new economic ideas about free markets and production. Adam Smith (1723–1790) made the well-known invisible hand argument: that the good of all would result from people working for their own benefit in a competitive market. Karl Marx (1818–1883), on the other hand, argued that the ability of factory owners to control the low wages of workers leads to excessive profit for the owners, called capitalists, which they then use to gain political power to further exploit labour. Marx explained how this imbalance led to boom-and-bust economic cycles, each of which consolidated more wealth and power in the hands of the powerful and increased the numbers of unemployed, thus putting downward pressure on wages and driving smaller firms out of business.

In the twentieth century, the use of fossil fuels generated much higher consumption, growing population and new ways to track transactions. The Great Depression and the projects of rebuilding Europe and Japan were the context for the thinking of John Maynard Keynes (1883–1946). Keynes was a central figure in the Bretton Woods Conference (1944), which established new economic order by radically changing the basis of world currency exchange. Keynes's theory is that labour and consumers are passive participants in the economy and that political economics should seek to focus wealth in the hands of the wealthy and businesses who will invest in growth, thus trickling down benefits to the masses by creating jobs. Keynes also advocated deficit spending by the federal government to maintain employment, which would avoid recession, and to stimulate economic growth. The Keynesian theory is that future growth will provide greater wealth to deal with deficit spending. Keynes also advocated taxes on unearned income (the dividends paid to shareholders or interest on savings).

Ecological and energy economics ideas proposed that ecological health or consumption of energy should be used to measure economic performance, with the aim of safeguarding sustainability. More recently, Elinor Ostrom (1933–2012) became the first woman to win the Nobel Prize in Economic Sciences for her analysis of the economics of governance of the commons. Ostrom proposed that the role of public choice in regards to decisions about how to exploit resources and ecosystems was important in developing governance and management practices to avoid ecosystem collapse or resource exhaustion.

There are dozens of theories in the history of political economics. Environmentalists, scientists and advocates for sobriety tend to disagree with the dominant Keynesian or neoliberal political economic theory that advocates for continued and continuous growth in production, productivity and consumption. The political economics theories usually assume that technological progress is a given. Political economic theory deals with the behaviour of consumers and businesses, but I have yet to see an economic theory that includes the behaviour of technologists and scientists.

6.4.1 THE EMPEROR'S NEW CLOTHES

An emperor loves clothes so much that he has a new and more glorious outfit made for each hour of the day. The economy of the kingdom has become dominated by the design and tailoring of clothes for the king and for his courtiers, who strive to emulate him. The import costs for silks and fine materials is very high, and the kingdom has run up a large deficit. The kingdom's warehouses are full of clothes the emperor and courtiers won't wear twice. The bridges, roads, farms and schools are falling apart. There is an environmental crisis in the farms outside the city, but the emperor will not hear any complaints and only wants to plan for more splendid clothing.

One day, two rogues arrive in the emperor's court and promise the emperor clothes made of cloth so fine that people who are not fit for their office will not be able to see it. Of course, the king must have these fabulous clothes. He pays huge sums, and the rogues go to work in a secret shop. The weaving takes a very long time, and the king sends his wisest and most trusted advisor to check on progress. Of course, the advisor sees there is no cloth on the loom, but he decides to report to the emperor that the cloth is indeed the most amazing he has ever seen.

When the garments are finally ready, the king makes a grand parade to display his new clothes to the people. Everyone can see that the king is naked, but it isn't until a child calls out 'He has no clothes!' that all the people break into laughter, because it *is* obvious. The emperor suspects the people are right, but stands even prouder, carrying on with the parade, and his courtiers hold the train of his robes higher even though there is really nothing there at all.

> **THE ROGUES HAVE RUN OFF WITH A FORTUNE, AND THERE IS STILL NO MONEY FOR NEEDED SERVICES.**
>
> *Based on: H C Anderson, The Emperor's New Clothes*

In the story about the emperor's new clothes, we see a familiar drama being played out, where the political economic leadership becomes obsessed with something that is glamorous and interesting but is, in reality, unimportant. This focus on trappings of wealth and power leads to underinvestment in infrastructure and services and the degradation of natural resources. It is interesting in the story that the common people go along with the illusion of wealth just for the spectacle. It is also interesting that the innocent child who has no economic or power interests is able to see the truth of the situation. In this story, what would you say is the role of the engineers?

You might like to think that our professional ethics would place us in the role of the truth-teller. But I submit that, since the energy and environmental crises of the 1970s, engineers, and, in particular, engineering researchers, have been playing the part of the rogues. Yes, the political economics might be off track, but if we technologists are participating in and profiting from that delusion of continuous growth and technological progress by providing new and more exciting energy solutions that aren't really there, then we are really the bad guys in the story.

Let's rewrite the story with the arrival in the kingdom of some transition engineers. They see the amazing clothing of the king and his courtiers and that many of the normal industries of a city have been taken over for the making fine clothing, wigs and shoes. They talk to the minister of economics, who explains that the GDP of the kingdom has been growing, along with the consumption of fine clothes by the upper class. They observe that the roads and sewer system need maintenance, the roof of the community hall is leaking, the schools don't have enough supplies, and the bridge into the city is at risk of collapse. They can see the evidence of overshoot in the economic activities: the king is spending more on clothes, but there is no return in benefit for that spending. The clothing-making sector is providing the demand, but it is not sustainable because the lack of investment in bridges, schools, and building and water system maintenance is undermining the ability to do business. The transition engineers propose a shift project for the kingdom. They will work with the guilds to set a cap on the production of fine clothing, which will decrease each month. The price of the fine clothes will be set at a high level and will include a renovation tax. The renovation tax will be used by the guilds and the mayor to repair and renovate the kingdom's infrastructure. The renovations can also include statues and memorials to the king and courtiers, for a price. The ego tax is used to repair the sewer.

7 Transition Economics
Balancing Costs and Benefits

Sustainability and adaptation are the two sides of survival: the balance between tradition and innovation, costs and benefits. Finding the balance requires risk management. If you never take any risks, then you will miss out on benefits. But if you take too many risks, the costs could very well be too high. Current economic modelling discounts both future values and costs, so it is no wonder that there is no way to manage the long-term risks of climate change and resource scarcity. Recent ideas of including values of ecosystem services and costs of pollution externalities in the conventional economics models have not changed the way risk is managed enough to change the emissions and consumption growth trajectories. A few economists have discovered that the free benefits of energy and resources are missing from their models, but how to change the equation to include the risks of boom and bust remains unsolved (Jackson 2009).

Transition economics analyzes costs and benefits based on real value. Real value is the same throughout time, regardless of the value of money. Real value can be converted to market value, for example, via mining and smelting hematite into iron, but this conversion consumes ore, energy and water. An hour of work by a skilled carpenter has real value, as does wood, iron and concrete. Construction converts this real value into essential real value like housing or manufacturing facilities, social value like schools or churches, or it can convert the resource real value into non-productive assets like casinos, convention centres, freeways or resorts. This may seem a contentious point. Casinos and hotels return profits, but they do *not* produce any real value. Highways and airports are used by travellers, but they do not create or produce real value. A prosperous society can afford large-scale entertainment and enabling infrastructure, for example, the Romans built colosseums, aqueducts and roads. But a sustainable society finds the balance between investment in essential and non-productive assets.

We need real value to produce food, clothing and shelter, and to provide essential services. If we irreversibly degrade or damage our real value endowment in the pursuit of currency, then we are risking our future prosperity. For example, farming produces food and fibre while consuming soil fertility, and using land, sun, water and labour. Industrial agriculture is aimed at high production of cash crops, but at the risk of unsustainable soil and water consumption and pollution. Since the invention of artificial fertilizers in the 1930s, gas and petroleum can also be converted to food, but continued industrialization has accelerated the degradation of soils and consumption of water in agriculture. We also need real social value in order to have a decent standard of living and equitable quality of life. Social elements like cultural traditions, schools and governments also have real value in producing cooperation, knowledge, and problem solving.

We have the computing power and analytical capabilities to examine the real value of free resources and humanmade systems at different times in the future. Our study of the problems of sustainability has clearly shown that we are in a time of transition. The past period of growth in use of energy and resources is transitioning to a future of down-shifting. We have seen that boom and bust has occurred in past transitions. However, the whole point of Transition Engineering is to accomplish a down-shift in fossil energy and resource use while uplifting real value through innovation, design and management. The forward operating environment (FOE) has hard constraints on free resources, and this one simple and obvious fact requires a whole new economic thinking (Raworth 2017). Thus, we will focus on sustained or improved real value and down-shift of energy use as the measure of merit for decision support, not net present value and not growth. The transition economics approach involves a perspective shift. We will move our frame of reference years into the future and construct the context for how real values at that time are affected by decisions made now.

7.1 BIOPHYSICAL ECONOMICS

Biophysical economics uses material and energy flows rather than prices to evaluate the performance of the economy. All essential activities require energy; thus, the key metric used in biophysical economics is energy return on energy invested (EROI). There are no physical constraints on the amount of money that can be created or used because money is no longer based on a precious metal standard. This makes it increasingly difficult to use cost alone to analyse energy systems. Biophysical economics uses energy, embodied energy and material flows for accounting rather than cash flows (King and Hall 2011; Dale et al. 2012a).

The Sankey energy flow diagram maps out how free primary energy resources are extracted from the environment, converted into consumer energy products (electricity and fuels) and distributed to end-use conversions (LLNL 2019). The entire point of having a fuel sector and an electricity sector is to provide energy to society for all of the end-uses. The basic model of the economy is a circulation of money being exchanged by producers for labour and capital, and money being exchanged by workers for goods and services that they consume. The electricity and fuels that flow through the economy all end up as waste heat and emissions after doing the work and providing the services. Energy and natural resources are essentially free and not accounted for in standard economic analysis (Jackson 2009).

7.1.1 BIOPHYSICAL ENERGY-RESOURCE-ECONOMY MODEL

Figure 7.1 shows the energy and material flows between the energy production sector and the economy (Dale et al. 2012b). The energy sector includes all of the electricity generation and fuel-processing industries (Brandt et al. 2013), where

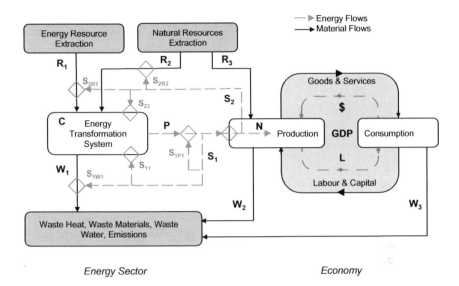

Energy Sector   *Economy*

FIGURE 7.1   The flows of energy and resources through the energy production sector and the net energy and resources to the production sector to provide the goods and services consumed in the economy actualized by the circulation of money.

C is the total energy conversion capacity. Energy transformation processes use primary energy resources, R_1, and natural resources, R_2, to produce consumer energy, P. Energy production requires electricity or fuels, S_1, and embodied energy in equipment and materials, S_2. The net energy to consumer in the economy, N, is thus the production from the energy sector minus the consumption by the energy sector. The energy transformation processes produce wastes that require management and removal, W_1. The total energy transformation consumes electricity and fuels for transmission and distribution, S_{1P1}; the energy transformation plant, S_{11}; the extraction processes that supply the energy sector, S_{1R1} and S_{1R2}; and the transport and disposal of waste from energy production, S_{1W1}:

$$S_1 = \sum S_{1i} + \sum S_{1Ri} + \sum S_{1Pi} + \sum S_{1Wi} \tag{7.1}$$

where
  S_1 = consumer energy purchased and used by the energy conversion industry
  S_{1Ri} = consumer energy used in extraction, processing and delivery of resources
  S_{1Pi} = consumer energy used in transmission and distribution of consumer energy
  S_{1Wi} = consumer energy used in removal, processing and disposal of energy sector waste

The total embodied energy in capital equipment used by the energy transformation system includes energy conversion hardware, S_{22}, and equipment used for resource extraction, S_{2R1}, S_{2R2}, for example, mining, diverting water, building roads to wind turbine sites:

$$S_2 = \sum S_{2i} + \sum S_{2Ri} + \sum S_{2Pi} + \sum S_{2Wi} \qquad (7.2)$$

where

S_2 = embodied energy (in materials and equipment) used by the energy conversion industry

S_{2Ri} = embodied energy used in extraction, processing and delivery of resources

S_{2Pi} = embodied energy used in transmission and distribution of consumer energy

S_{2Wi} = embodied energy used in removal, processing and disposal of energy sector waste

Table 7.1 gives energy and resource conversion and consumption analysis parameters using the variables defined in Figure 7.1 and gross domestic product (GDP). GDP is a monetary value of transactions in an economy and is defined and calculated variously by government agencies. GDP counts all expenditures as positive whether they derived from purchasing goods or providing healthcare to people affected by pollution, remediation of mining environmental damage or cleaning up oil spills, so it is not an accurate measure of social well-being.

The parameters in Table 7.1 are referred to as macro-economic because they are used to make general comparisons between different development options. The whole economy of a country is more prosperous if all these factors are more positive, for example, low resource intensity, high efficiency, low external costs.

TABLE 7.1

Assessment of Energy and Resource Characteristics of the Whole Economy, a Particular Sector or a Particular Subsystem Using the Energy-Economic Biophysical Model in Figure 7.1

Indicator	Parameters	Example
Energy conversion efficiency	$\eta_C = P/R_1$	MWh$_e$/ton coal
Second law efficiency	η_C/η_{Carnot}	Engineering quality of a plant
Energy resource intensity	P/R_2	Gal ethanol/gal water
Energy intensity of the economy	P/GDP	kWh/\$
EROI	$EROI = P/(S_1 + S_2)$	MJ/MJ over a period of time
	$EROI = \dot{P}/(\dot{S_1} + \dot{S_2})$	MW/MW at a point in time
Embodied energy	R_1/GDP	MJ/kg steel
Net energy to the economy	$N = P - (S_1 + S_2)$	GWh electricity or GJ fuels
Utilization factor	$U = 8760 \times C/N$	Energy sector productivity
Externality loading of conversion	W_1/P	Ton-CO_2/MWh
Lifecycle assessment	$\sum R$/GDP	Resources/unit good or service
External costs	$\sum W$/GDP	Impacts/unit good or service
Impact footprint	W/GDP	Carbon footprint
Levelized cost of energy	$LCOE = \sum S + \sum R + W_1$	Annualized investment cost
Energy prosperity	$N/P = 1 - 1/EROI$	Rate of return on energy

Some of these parameters have the same names and basically the same meaning as in engineering of the energy systems. For example, the first law energy conversion efficiency is defined as what you get compared to what you put in, and it is used for both energy generation and end use appliances.

The second law of thermodynamics has been stated different ways: water can't flow uphill without energy input, chemical reactions cannot proceed backwards without energy input, material cannot be concentrated without energy input, and energy can't be converted from one form to another without losses. The second law expression for energy conversions is the following: it is impossible to have a conversion process that is more efficient than a reversible process, and it is impossible to have a thermal power cycle more efficient than a Carnot cycle. The reversible model is an imaginary process that has no losses, no friction, no pressure losses, and so on. The second law efficiency is used by engineers in the development of technology and in system design and selection of components. The concept of the Carnot efficiency can be used to inform policy and investment decisions to a certain degree. The Carnot efficiency is a reversible model for energy conversion by an ideal thermodynamic cycle that transforms heat supplied to the engine into work:

$$\eta_{carnot} = 1 - (T_C / T_H) \tag{7.3}$$

where

T_C is the environmental temperature (usually 293 K)

T_H is the energy resource temperature (in Kelvin)

Nuclear energy has the highest resource temperature, and coal and gas combustion temperatures are higher than wood combustion, which in turn is higher than nearly all geothermal temperatures. Solar thermal temperatures can be as high as 600°C for a parabolic dish concentrator, and around 250°C for a tracking parabolic trough concentrating collector. The Carnot efficiency for high-temperature resources is necessarily higher than for low-temperature resources. The second law efficiency is the actual energy conversion performance for a plant compared to the Carnot ideal model of the plant. The thermal efficiency is the useful work produced by the thermodynamic cycle compared to the heat transfer into the cycle.

The second law efficiency is the actual power output compared to the 'ideal' *reversible isentropic* model of the same component or system. Recall that the reversible isentropic model is an ideal construct where friction and loss do not occur, and thus it is an impossible ideal but useful for comparison. The isentropic efficiency of modern gas turbines, operating at their optimal speed, pressure ratio and mass flow rate, can be as high as 95% of the ideal power output. However, if the engine is run at conditions different from this operating state, or if the turbine blades become eroded or dirty, the isentropic efficiency can be much lower. Economic analysis of energy conversions often assumes optimal performance, so it is important, particularly when considering intermittent energy sources like solar and wind, that we recognize that much of the performance will be far away from optimal.

7.1.2 Energy Return on Energy Invested (EROI)

EROI is an important concept for sustainable energy engineers to understand and to communicate to the public and policy makers (Ballard et al. 1978; Spreng 1988). In terms of Figure 7.1, the EROI is:

$$\text{EROI} = P/(S_1 + S_2) \tag{7.4}$$

where
    P = rate of energy production (kWh/analysis period)
    S_1 = consumer energy used in energy production (kWh/analysis period)
    S_2 = embodied energy in energy production capital equipment (kWh/analysis period)

EROI can be evaluated for the whole energy sector in a given year (Dale et al. 2011), or it can be assessed for a particular energy production technology over the useful life of the plant. The only reason for building an energy conversion plant (investing embodied energy in capital equipment, S_2) and operating the plant (investing process energy, S_1) is to produce consumer energy for the market (distributing and selling consumer energy, P). EROI is a measure of the energy profitability of an energy transformation system. The energy invested in the transformation system cannot be used by the economy for other activities. EROI is also an indication of the surplus energy available for the economy to meet current demand, to provide maintenance and replacement, and to supply manufacturing and new construction.

Example 7.1

Let's explore the concept with an example of making shoes. EROI is a type of input/output (I/O) analysis but specifically for the energy sector. We might not worry too much about using energy to produce shoes. People need shoes, and the energy used in manufacturing becomes part of the consumption in the economy. But what if shoes were needed in order to produce shoes? If you had to buy a pair of shoes in order to produce a pair of shoes, what would be the point of making the shoes? You could not make a profit unless you could convert bad shoes into good shoes that someone was willing to pay a very high price for. But while upgrading for a willing market may be profitable, you have only produced an exchange; you have not provided shoes. If the shoe return on shoes invested in 1:1, then the shoe-manufacturing sector is using resources, possibly making money, but not providing the economy with shoes. I use the example of shoes here because I assume that readers do not have any particular emotional attachment to shoes and that they understand that people can be induced to pay ridiculous prices for shoes due to motivations of fashion. Our economy demands shoes because they are useful. We consume them at a certain rate and thus need a supply of new shoes in order to carry out our essential activities. If the government subsidizes the conversion of some of the shoe manufacturing capacity to shoes that require a functional shoe from us in order to supply a fashionable shoe, then the net shoe production from this sector would be zero and the effect would be loss of shoe supply capacity.

Yes, this shoe example is a metaphor for green energy alternatives with low EROI like ethanol, solar PV with low utilization factor, hydrogen, carbon capture and storage, and solar PV with battery storage. Politicians are comfortable using public funds for research in green alternatives. A carbon footprint analysis may show a benefit of using the green energy alternative. It is physically possible to build the alternative technology and, with sufficient subsidies, it might be profitable to manufacture and install them. It might also be fashionable for consumers to purchase the green energy alternative technologies. But if the EROI is low, this whole exercise is not an energy supply strategy; it is simply a consumption of resources and ultimately a drag on the economy.

The concept of EROI is simple, but exact calculation can be difficult. Companies are not likely to invest in an energy platform with low EROI because it would be bad business. However, governments have invested in research into low-EROI concepts, which generates interest and favourable public opinion. This popular attraction of low-EROI alternative technologies in turn influences policy that incentivizes energy platforms with poor EROI. As more resources from the economy are invested in low-EROI technologies, the overall EROI of the economy declines, increasing the risk exposure to supply security issues, and ultimately increasing carbon emissions as more inputs from fossil resources are needed to realize the smaller renewable production.

EROI lets us compare different energy transformation platforms and make informed investment decisions. The objective of investing is to realize a profit. In the energy system, the 'profit' for the society as a whole is the net energy that can be used by the economy. The net energy yield, N, is given by:

$$N = P - (S_1 + S_2) = P(1 - 1/EROI) \tag{7.5}$$

7.1.3 EROI Analysis

Figure 7.2 illustrates the EROI calculation method for a renewable resource. Energy production is positive, and energy investment is negative. Energy production accumulates over the lifetime of the power plant and is plotted in each year along the x-axis. The construction requires energy input and uses embedded energy in concrete, steel, aluminium in plant and equipment. Once the plant is built, it produces energy. We usually assume the same production rate each year unless we know the derating factor. There are fuel inputs, operation and maintenance that are negative. At the end of the plant's lifetime, there will be decommissioning and recycling energy costs. Clearly, the EROI and net energy yield are sensitive to the plant lifetime. It is also easy to see how an EROI = 1 would not be an energy production system as the plant would only produce the same amount of energy that went into creating and running it.

The net energy yield will be negative at the early stage of a power plant. The point at which the cumulative net energy return reaches zero is the energy payback period (EPP). Coal and hydropower plants and oil fuel production have extremely short EPPs, and so EROI is usually discussed on an annual basis, whereas wind or solar EROI usually refers to the lifetime of the plant. Biofuels are derived from crops, so they are usually analyzed on an annual energy flow basis.

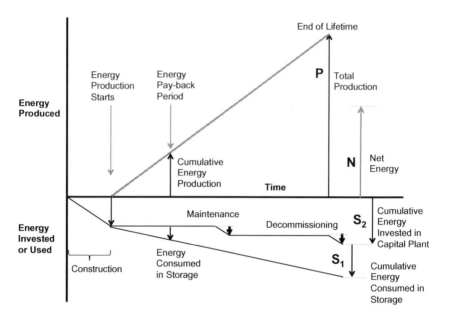

FIGURE 7.2 Energy input and output for a renewable energy generation system during its lifetime.

In electricity generation from thermal power plants, rejected thermal energy and internal parasitic power uses for pumps are not counted in the input energy, S_1, because the production, P, is the gross power output of the plant. Petroleum exploration, drilling and oil well equipment are included in S_2, and the electricity for pumps and fans used in refining the petroleum into consumer fuel products is counted in S_1 if the electricity is purchased from the grid and not produced on-site from parasitic consumption of some of the fuel produced. Parasitic power or fuel consumption internal to the energy conversion industry did not exist as a marketable product that could be used in other parts of the economy, so it is not counted in S_1. For conventional electricity generation, EROI is vastly different depending on whether the energy input from the fuel consumed is considered as a free resource taken from the environment, like solar or wind energy, rather than a consumer fuel taken from the economy. There is currently no standardized approach for calculating EROI, so all reported values should be examined closely to ascertain the underlying assumptions (Murphy et al. 2011).

The diesel fuel for mining and electricity for processing and handling the fuel is usually a small energy input for a coal power plant compared to the power produced. The embodied energy in the plant and maintenance are energy inputs. The biggest difference in reported values for fossil fuel generation and renewables is that the fossil EROI is often calculated on an annual basis, whereas the EROI for wind and solar are determined for the lifetime generation. EROI for coal power generation is in the range of 10.

For nuclear power plants, EROI would depend greatly on whether the energy needed for the mining and clean-up of mining and processing sites, storage of spent fuel and decommissioning is counted. Nuclear power plants are vastly energy intensive to build, but EROI, not counting the uranium processing and waste storage, is as high as coal. Much of the historical mining and processing of uranium in the United States had a subsidy from nuclear weapons investments. The net energy return for future nuclear developments would need to include the fossil fuel used in the extraction and environmental management associated with the industry as well as some accounting for long-term waste storage as these are now much more visible costs to society.

A key difference between EROI for fossil fuel or nuclear-generating systems and renewable technologies, such as solar or wind, is the utilization factor, U. EROI for conservation measures and waste heat recovery is calculated using the same approach as for energy transformation systems. The energy savings is accounted as the energy production, P, because the reduced demand for consumer energy due to conservation is essentially made available for use somewhere else in the economy.

Example 7.2

EROI of a renewable power plant with and without battery storage is illustrated in Figure E7.2. During the construction phase, two units of energy are invested to build the system. The energy system produces one unit of energy every year for 12 years. During its operational life, one unit of energy is invested for maintenance, and one unit of energy is needed to decommission the system. The cumulative lifetime energy produced is $P = 12$ units.

Thus the EROI for this example over the lifetime is:
$EROI = P/(S_1 + S_2) = 12/(2 + 2) = 3$
The net energy available for consumption is:
$N = P - (S_1 + S_2) = 12 - 4 = 8$ units

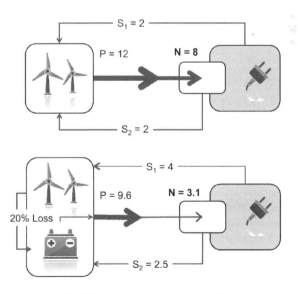

FIGURE E7.2   Example of EROI for wind power with and without battery storage.

Now consider if this renewable energy system is used with battery storage. The amount of energy produced that can be consumed to meet demand would be reduced by the 20% round-trip losses, so $P = 9.6$ units. The batteries require one unit of energy input to manufacture, need to be replaced at year 6 and require 0.5 units of energy to recycle both sets of batteries used:

$EROI = 9.6/(4 + 2 + 0.5) = 1.47$

The net energy is:

$N = 9.6 - 6.5 = 3.1$ units

7.1.4 EROI AND PROSPERITY

EROI is a determining factor for prosperity. The astute reader will notice that we have not yet discussed the price of the energy or resources in the biophysical analysis. Prosperity depends on the production first and the prices second. Consider a farm system. The farm purchases goods and energy and produces food. Survival is achieved if the food production is equal to the food needs of the farmers and farm animals, but the farm will be declining as there is no surplus production to take to the market and thus no way to pay for repair and replacement of implements. If there is no surplus to take to the market, it does not matter what the price in the market is. Subsistence is achieved if the food production meets the farmer's needs and provides enough surpluses for sale to pay for the goods and energy needed to maintain production. In the subsistence case, the prices are important because the value of the crops sold needs to balance with the price of the goods needed to maintain production. If there is inflation in farm equipment and input prices, but deflation of commodity prices, then farmers can find they are in a declining position. They could work harder and lay off farm hands to get back into balance of income and costs, but their position is still declining. This situation of declining prices for agricultural commodities has been an issue since World War II (Sumner 2009).

Prosperity is possible if there is substantial surplus, and no inflation of the cost of inputs. A farm in decline would have to go into debt in order to operate, and there would be no spare cash to paint the barn or fix the farm gate. A subsistence farm could avoid debt and keep the equipment running, but it could not afford to paint the house or purchase consumer goods. A prosperous farm would have well-maintained buildings, the latest equipment, well-paid labourers, healthy animals, educated children, some luxury goods and so on. If all of the primary production industries are prosperous, then their consumption of goods and energy and employment of workers provides markets for others, and we have a prosperous economy.

This biophysical model of prosperity has always been true in every society. The anthropogenic system dynamics model has economic exchange as the actuator. People exchange money for goods and are paid for their labour. But money is not goods or labour. Prosperity depends on production being larger than inputs, and little or no inflation in the price of the primary products.

In fact, there is an inverse relationship between prosperity and the price of commodities – scarcity causes prices to rise for primary goods, but it does not result in prosperity, as demonstrated in many historical cases, including the energy crisis caused by the Organization of Petroleum Exporting Countries (OPEC) oil embargo (Erten and Ocampo 2013).

In a business, the profit margin is defined as the net profit divided by the total revenues from sales. The profit margin is a measure of the performance of the business. The optimal profit margin would be 100%, meaning that all of the revenues were pure profit, and there were no costs of doing business – but this is not realistic. A good profit margin for a business is around 25%, and a profit margin below 10% is unsustainable, as the profits are not sufficient to maintain the depreciating assets. Depreciating assets are the buildings and equipment used in the business. Assets need to be maintained and replaced at the end of life. Even not-for-profit organizations must have a profit margin above 3% in order to be sustainable. A bookstore can analyse the profit margin on different kinds of books it sells, and the manager would likely decide to increase the floor space dedicated to books with higher profit margins. Books that cost more to source and shelve than they return in sales would be discontinued. The bookstore can borrow funds to build an extension, but then its costs of business would be higher. Its manager can hope that the revenues will be higher, but there is a risk that the profit margin will not be higher. Often, a smaller revenue stream with a higher profit margin is a better business position.

All of these business ideas apply to the biophysical system as well. The net energy return to the economy as a percentage of the total energy produced is a measure of the energy prosperity of an economy, and effectively an energy profit margin:

$$N/P = 1 - 1/\text{EROI} \qquad (7.6)$$

Figure 7.3 is a plot of the energy profit margin N/P versus EROI and gives a range for prosperity, subsistence and decline. An N/P of 95% would mean that generation

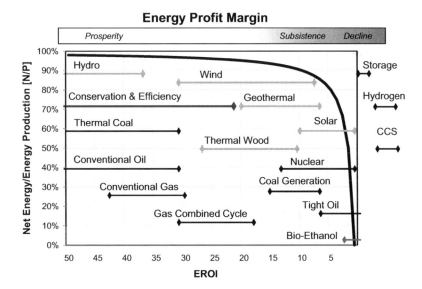

FIGURE 7.3 Energy profit margin is the ratio of net energy to the rate of energy production (N/P), and depends on EROI. The EROI range of various energy options are indicated.

assets capable of producing 100 units of energy provided 95 units of energy to the economy. EROI is 20, which means that, for every 100 units of production, five units of consumer energy and embodied energy are used by the energy sector. Energy would be abundant in this case, even if the total amount of production was low. There would be surplus energy for use by all other sectors of the economy because the energy production sector consumes only a small proportion.

An N/P of 50% means that the energy sector with generation assets of 100 units provided only 50 units to the economy. EROI is 2. Total production capacity investment for an energy conversion system with $N/P = 50\%$ would have to be twice as large as that for the $N/P = 95\%$ system in order to provide the economy with the same energy supply for the non-energy sectors. Figure 7.3 shows that, once EROI drops below 5, the energy profit margin decreases precipitously, meaning that the society is investing huge resources in the energy sector but getting a lower return on the investment.

Another way to think of the N/P ratio is as an energy tax. Taxes are a highly political issue, but the biophysical interpretation has some relation to the arguments around the role of taxation in society. If no taxes are collected, then there can be no public investment in infrastructure, education, regulation or services. Most people can appreciate that the after-tax wages and revenues are what keeps the economy working as people and businesses can afford to meet essential needs and have disposable income. Therefore, there is a balance between sufficient tax to provide the public infrastructure and services without collecting too much tax and dragging down economic activity. The percentage N/P can be seen as the 'take-home pay' in biophysical economics. The society with $N/P = 95\%$ is only paying an energy tax of 5%. A society with $N/P = 50\%$ has to give half of the total production from the energy sector back to the energy sector. Let us be very clear, EROI = 1 is not energy production at all. An EROI = 1 would mean that consumer energy that could have been used for other purposes and essential activities was used by the energy sector, but there was no return on that investment.

7.1.5   EROI of Biofuel

Production of ethanol biofuel from wood cellulose or algae has been shown in multiple studies to have EROI around 1, with an energy profit margin of less than 5%, or an energy tax of more than 95%. Figure 7.4 illustrates the EROI calculation for using the corn grown in New Zealand in 2004 as the feedstock. The EROI = 1/ (0.03 + 0.04 + 0.54) = 1.6 if the 2.32 MJ of the food energy value in the crop is not counted as an input to the fuel and is counted as a free resource. However, we don't believe that farmers would count the corn as a free resource because it has food value. If we count the corn energy value, the EROI = 0.34 (Krumdieck and Page 2013).

Growing the biomass, building the plant and using natural gas and electricity produces a fuel with energy value roughly equal to the fuel, gas and electricity consumed by the ethanol industry. All of the energy inputs could have been used for transportation rather than making ethanol. Thus, the ethanol industry does not accomplish anything in the biophysical economy. In the financial economy, monetary investments and costs are incurred in all steps of the ethanol industry, and subsidies were given to

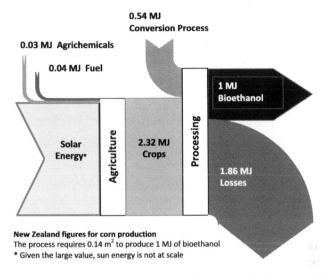

FIGURE 7.4 Energy inputs for agriculture and ethanol processing for New Zealand corn. (From Krumdieck, S., and Page, S., *Energy Policy*, 62, 363–371, 2013.)

build the factories and provide water. Thus, there was increased GDP because there was increased spending, but no net energy benefit was achieved. This is a very bad deal. That same financial investment used to shift fossil-fuel truck freight movement to rail by building rail infrastructure and innovating new rail technologies and business models would reduce the consumption of a finite, high-EROI fuel that could have been used in the economy for essential activities at all times in the future. And building rail infrastructure would also provide jobs and increase GDP.

Use of batteries, or production of hydrogen in an energy system, consumes energy in the conversion from electricity to chemical potential energy in order to store energy for later use. Energy storage always has EROI less than 1 and represents a significant energy cost of doing business. If the energy storage is added to an energy production system with an already low EROI, like solar PV, then the possible prosperity for the economy using that system will be low. It is fundamentally impossible to run an economy on hydrogen, solar PV with battery storage, or liquid biofuels. The proposed technology of carbon capture and storage (CCS) for coal-fired power generation similarly is an energy penalty on the energy production system and would drop EROI for coal-produced electricity from the range of 10 to below 5 (Page et al. 2009).

7.1.6 TIME DEPENDENCE OF EROI

EROI changes over time. When a resource is first developed, the markets and end-uses are not established and production is usually not very efficient. Technology improves rapidly as the market grows; EROI increases at first but eventually depletion effects start to erode EROI (Dale et al. 2011b). EROI of oil production has declined from EROI = 50 over the past 70 years. The most productive resources were initially used, but once they peaked and started to decline, it was necessary to drill deeper for increasingly smaller less permeable oil fields. It is estimated that, by 1970, EROI of oil production had dropped to 25. The EROI was somewhere between 10 and 18 in 1990. Today, EROI of new oil production is somewhere between 3 and 4 (Cleveland 2005) as the resource mix includes smaller reservoirs and tighter formations. This decline in EROI is the result of having used up the most productive sources, such as the Spindle Top in Texas, and about half of the Ghawar Field in Saudi Arabia, which were less than 500 m below the surface. In contrast, the fields in the Gulf of Mexico are more than 5,000 m below the surface and hold only a fraction of the Saudi Arabian and Texas oil fields (Cleveland and O'Connor 2010).

Tar sands and shale resources hold an enormous amount of hydrocarbon, but the energy inputs and financial cost of extracting and processing these hydrocarbons into consumer energy is many times greater than that of the oil fields of the past. It has been estimated that, to extract oil from tar sands, EROI is less than 5, while for oil shale, it may be as low as 2. This means that an enormous amount of energy needs to be invested for a relatively small amount in return (Gupta and Hall 2011). In addition, extraction of oil from both tar sands and oil shale has a deleterious effect on the ecology, as well as requiring large amounts of water.

7.1.7   EROI of Down-Shift Projects

Conservation and efficiency improvement are down-shift projects that decrease consumer energy consumption. Projects like rebuilding or modifying buildings require energy input, but they return energy reduction. We count the energy savings as the energy production in the case of conservation measures. Energy conservation through behaviour modification and shift to low-energy systems can often have long-term N/P near 100 (Krumdieck and Page 2013). The biophysical engineering analysis can provide decision support with a different perspective to understand the profound difference in benefits between investments in different energy transition options.

EROI for wind- and solar-generated electricity has been increasing over the past decades as technology, efficiency and lifetime have improved. EROI for improved efficiency has actually declined with time. The first wave of energy conservation and energy efficiency improvements was made in response to the energy crisis of the late 1970s. The low standard of appliance efficiency and building thermal performance at the time meant that even moderate added expenditures in energy management could reduce energy use dramatically. Today's light-emitting diode (LED) lighting, appliances and building construction have low energy intensity, and new investment in marginally better energy efficiency will provide less energy savings than in the 1970s. In addition, buildings built before the higher standards were put into effect are often not worth the expense of retrofitting to conserve energy. All energy and resource development is subject to the 'low hanging fruit' scenario. We always use the resources that are easiest to reach first.

7.1.8   Energy Transition, Sustainability and Prosperity

Using the biophysical perspective, the energy prosperity of a society depends on having a high EROI and having enough energy to meet consumer and business needs, as well as upkeep of the infrastructure and built environment. Figure 7.5 illustrates the prosperity difference between high and low EROI energy sectors, as well as high- and low energy consumption from the economy's point of view.

The energy transition is from high-energy intensity to low-energy intensity and to very low consumption of fossil fuels limited to essential services. Prosperity can be achieved with low energy use as long as the supply is sufficient for essential activities, maintenance and replacement. Energy end-use systems will be redeveloped through successful transition innovations and management to require very low energy to provide essential services. Transition Engineering of these complex systems is essential, and the shift projects should have relevance today and 100 years in the future.

The largest share of net energy is used to carry out personal, business, education and service activities. This would include the bulk of energy used in transport and operation of buildings. Primary production and manufacturing use energy to produce the food, materials and goods for the economy. The essential production for a society uses at least 10% of the net energy. Maintenance of buildings and infrastructure is essential for prosperity, and uses at least 10% of the net energy. Equipment and buildings that are replaced at the end of their useful lives or that are being redeveloped for the energy shift would require around 4% of the net energy.

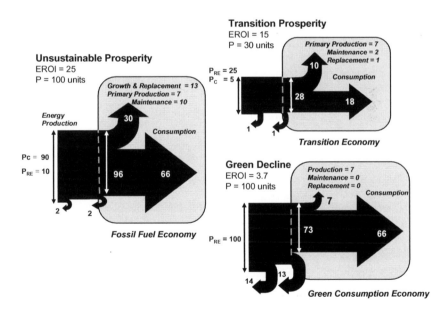

FIGURE 7.5   Net energy to the economy for three energy transformation systems, with energy production from fossil, P_C, and renewable, P_{RE}, resources. Unsustainable prosperity is representative of the energy system in the 1970s; green decline has low EROI and high consumption. Transition prosperity needs 1/3 the energy of the unsustainable system for consumption and is sustainable because the maintenance and replacement needs are met.

The Unsustainable Prosperity model in Figure 7.5 is reflective of the fossil fuel economy around the year 2000. For every 100 units of energy production, 10% is renewables, mostly hydro. Overall, the energy transformation sector has EROI = 25, so the net energy is 96 units. The net energy production supports large end-use consumption as well as growth and the necessary primary production and maintenance. However, the fossil fuel economy is not sustainable due to environmental constraints.

The Green Decline model explores the concept of substitution of fossil fuels with renewable wind, solar and biofuel with storage. This model is reflective of the current policies and social narrative for building more renewables. The Green Decline scenario shown in Figure 7.5 assumes that the consumption is roughly the same as the fossil fuel economy. The total production is the same as the fossil fuel model, but the net energy to the economy is almost 25% lower because of the lower EROI of renewable energy with storage. If the consumption is not reduced, then the economy will inevitably be declining, as there is no energy for maintenance or replacement. Also, recall that the low utilization factors of wind and solar mean that in order to achieve energy production of 100 units of wind, at least 333 units of capacity would have to be installed. In order to produce 100 units of solar, 555 units of capacity would have to be installed.

The Transition Prosperity model in Figure 7.5 meets the COP21 requirement of 80% reduction in fossil fuel and uses primarily high EROI renewables. It has 1/3 lower total production than the unsustainable fossil fuel economy. The Transition Prosperity system meets the requirements for prosperity, in that production is sufficient for the population and maintenance and replacement are 10% of net energy. The key lesson of the Transition Prosperity model is that, if the end-use consumption of the society is reduced by about 70%, then the society could have a prosperous and sustainable economy. Much of the world's hydropower was developed last century. Large hydropower plants often have utilization factor above 0.9 and EROI over 80%. Thus, the hydro that countries already have is the foundation for their energy prosperity and can be taken as the target production level for the energy transition.

7.1.9   EROI of Renewable Energy

EROI of renewable energy sources varies widely. As shown elsewhere in this book, EROI of modern wind systems in a high-quality location can be over 20, and solar thermal energy, depending on its location and eventual use, can be 7–25. On the other hand, EROI for ethanol from corn is only barely greater than unity, which means that it is not a sustainable energy source. Ethanol from corn has received substantial financial support from the US government for its production. This is a result of politics, since the corn-growing states represent an important political power in the US Congress. The EROI of ethanol from sugar cane in Brazil may be as high as 6–8. Considerable research worldwide seeks to produce ethanol from cellulosic material and algae, and the technology is not sufficiently advanced to make quantitative estimates. But the numerous processing steps and separation processes are more energy intensive than for high sugar crops.

Hydroelectricity has the highest EROI of any power generation platform, from 40 to 100 depending on whether a large dam and reservoir are needed, and the available head and flow. For example, some hydroelectric facilities can be built at the natural outlet of a pre-existing lake. Most of the hydroelectric dams in the United States were built many decades ago using high-EROI diesel fuel (50+) and coal for making the cement and steel (EROI approximately 80). The water flowing through hydro turbines does not need to be extracted from mines and transported to the power plant; it arrives courtesy of climate and weather, geography and gravity. In areas where hydro generation was built, prosperous economies based on plentiful, secure, on-demand electricity quickly developed. Examples include Grenoble, France, and Milano, Italy. Pumped hydro storage has been used in a few places like the New Waddell Dam in Arizona, where afternoon peak demands are many times higher than night-time demand, and a nuclear power plant provides a large and non-interruptible baseload supply. The investment in construction of an artificial lake and dam, and the redundant power generation system for pumped storage do not produce any *new* electricity.

Thin-film PV modules use very little semiconductor material (NREL 2004). The major energy inputs for manufacturing are the substrate on which the thin films are deposited, the film-deposition process, and facility operation. Substantial energy is used in mining and processing rare earth minerals used in wind generators and solar PV. National Renewable Energy Lab (NREL) reported that it takes 120 kWh/m^2 to make frameless, amorphous-silicon PV modules. Adding another 120 kWh/m^2 for a frame and support structure for a rooftop-mounted, grid-connected system, assuming 6% conversion efficiency (standard conditions) and 1,700 kWh/m^2 per year of available sunlight energy, the EPP is about three years and EROI = 11.75, assuming a 30-year life, but EROI = 7.8 if the lifetime is 20 years.

Wind-generated electricity is the fastest growing renewable generation. Utility-scale wind farms were first built in the late 1970s. Technology and wind farm design have improved greatly since then. The average EROI for systems built in 1983 was about 2.5, while for systems built in 1998 and 1999, the average was about 23, a nine-fold increase. Overall, the EROI for utility-scale wind power can be estimated to be around 25 (Kubiszewski 2010). If the wind farm were located at a place where new transmission lines would have to be built to deliver the power to a user, additional energy input for grid extension could be needed. Small wind turbines under 30 kW have EROI less than 10. If the wind were to be stored, then the energy embodied in batteries would need to be included. Homes that are off-grid can meet their essential needs with small-scale wind and battery storage, but there is no surplus for production, maintenance or replacement. You can meet some residential needs with wind and storage, but you can't run a wind turbine factory.

7.1.10 Net Energy Analysis and Future Scenarios

Future scenarios are an important part of the Transition Engineering method. In Step 3 of the seven-step method outlined in Chapter 4, we explored how future growth of new energy technologies could affect the energy prosperity of society. The basic method is the if-then approach of setting up the assumptions and modelling the results. The business-as-usual (BAU) scenario proposes that historical trends in growth of energy production and demand using fossil fuels plus recent trends in rapid expansion of wind and solar and research into other alternatives will continue. You can easily search the internet for future BAU energy scenarios based on growing future demand from the International Energy Agency (IEA), British Petroleum (BP), Shell, Energy Information Administration (EIA) and many others.

Here we will look at the BAU scenario from 2012, which used the biophysical economics and EROI of energy production (Dale et al. 2012b). In this scenario, the availability of free energy resources is assumed to be the ultimately recoverable reserves of fossil energy and the ultimate technical potential of renewable resources available for human use – that is, every possible resource will be developed for energy whether or not it is currently protected habitat or producing food. The BAU energy scenario using the biophysical model includes the following:

- Current conventional oil production is at peak and continues to be produced at the highest possible rates during well depletion through enhanced oil recovery methods.
- The tar sands, tight shale oil and gas, deep sea oil, Arctic oil and gas and all known reserves are developed. Demand continues to equal supply and does not drop below supply. Production must be increased at a higher rate over time because of lower EROI of the new petroleum sources.
- Hydro currently in place will be maintained and refurbished, and historical EROI will continue into the future.
- Nuclear currently in place will be retired at end of life and waste will be maintained safely.
- Current biofuel production is maintained and replaced after 20 years.
- Current geothermal production is maintained and further developments are made to achieve the stated targets in the countries that have geothermal resources.
- Wind and solar continue at the recent growth rates with the current EROI being the maximum achievable. The new higher EROI wind and solar assets increase the fleet EROI over the next 15 years until all older, low-EROI installations have been decommissioned. After 2040 the proportion of production of turbines and PV panels that is used to replace end-of-life units starts to grow at the same rate as the historical deployment grew, and the growth in capacity levels off.
- We do not assume depletion of any rare earth elements or other materials in this scenario. Lifetime of all turbines and panels is assumed to be 20 years.
- Solar thermal energy conversion (STEC) is maintained and grows to the estimated maximum capacity, as limited by suitable sites and water availability.
- The currently estimated future availability of other alternative energy production technologies are used as the entry point, and the entry-level EROI from research papers is assumed to increase to the theoretical maximum over time, then to decline. These assumptions are the most speculative and involve technologies that are not currently commercially manufactured, such as tidal power, ocean thermal energy conversion (OTEC) and wave energy.

- Hydrogen energy storage, large-scale battery energy storage and CCS are not included in this BAU scenario.
- The cost of energy was not used in the analysis to change the mix or the development rate. This BAU scenario is the most aggressive pursuit of growth of *all* energy resources technically achievable except nuclear. This scenario would result in greenhouse gas (GHG) forcing of more than 8.6 W/m^2, and global warming of well over 11°C, and more than 70 m sea level rise.
- The scenario only models supply, not demand. It is assumed that whatever energy is produced and brought to market will be purchased and utilized.

Figure 7.6 shows the result of a biophysical system analysis for the BAU scenario. The actual data for past production and EROI were used to calibrate the Vensim model used. The total energy produced and the net energy delivered to the economy are very close in the results of the historical model. The future projection of the peak production of fossil fuels reaching nearly 600 EJ/yr by 2040 also agrees with other published scenarios (Wang et al. 2017). The historical EROI for the whole energy sector is over 20, until 2010 so $N \cong P$ in that timeframe.

The total energy production continues to increase at the historical rate even as conventional oil declines due to a surge in renewable capacity and development of non-conventional oil and gas reserves. However, EROI of these resources is low, and the fleet average EROI starts to decline. Thus, the net energy yield has begun to diverge from the total energy production for 2010. The increasing decline of EROI

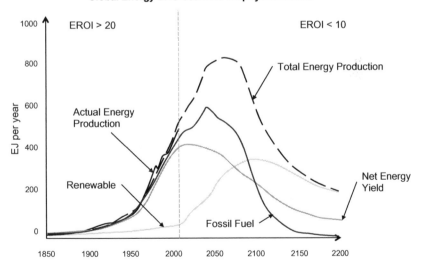

FIGURE 7.6 Net energy analysis of global total primary energy, assuming aggressive development of all energy resources and using the time-dependent EROI of the resources, shows the transition from fossil fuels to renewable energy by about 2150, but it results in a decline of net energy to end-users over the current century. (From Dale, M., *Ecol. Econ.*, 73, 158–167, 2012b.)

causes decline of net energy for all non–energy sector uses despite greatly increased total fossil fuel and then renewable energy production. Even though renewable energy production is expanded to technically possible, although unlikely, levels, the energy for use by society for all of its activities continues to decline. It is not clear that a steady state for renewable energy production exists as long as the low EROI conversion technologies are in the mix. If the energy sector is consuming two-thirds of the energy produced, then in fact more net energy could be returned to the economy by shutting down the low-EROI operations. This 'growth in production at all costs' BAU scenario exhibits a clear peak and decline of net energy for society.

This analysis raises some interesting questions. If the net energy to the world's economy is already peaking, then how is that being noted and felt? The net energy represents the consumer fuels and electricity that are made available in the market for non-energy sector uses. Although the total energy production could continue to increase for a time, a larger amount of that energy is used in the energy transformation industries for lower EROI resources. This can be seen currently by analyzing sector consumption data, which shows rapidly increasing consumption for the energy production sector.

Running out of fossil fuels is not the problem. The remaining fossil fuel reserves are larger than the amount of fuel used to date. The problem is that most public and private investments assume continued demand growth. Fossil fuels will always be the highest energy density resource, and they will always be in demand by industry, military, transport and domestic services, even at extremely high prices. High-EROI energy resources are required for prosperity. The mission of energy Transition Engineering is to drastically reduce the production and use of all fossil fuels in order to preserve a secure supply of these fuels into the future for essential needs, at a much lower level. This is why every shift project has the explicit requirement of reducing fossil fuel use by at least 80%.

TRANSITION ENGINEERING CONCEPT 6

Without hope, we give up. Without a realistic plan, we are lost. Therefore, we only need two things.

7.2   TRANSITION ECONOMICS AND FINANCIAL ANALYSIS

The key point of difference between conventional financial analysis and transition economics is the shifting of perspective in time. The time value of money used in conventional economics assumes that the economy will continue to grow, inflation will occur, returns on investments will increase revenues, and essentially this will all make the value of money in the future lower from today's perspective. The nature of the energy transition necessarily means that growth in the future, in the same sense as in the past, is not likely. The interdisciplinary transition innovation, engineering and management (InTIME) project economic analysis and decision support is based on the energy and materials performance of the business or organization at every point in time, not just the net present value of cost savings.

The InTIME methodology described in Chapter 4 is used to develop novel and creative ideas for shift projects. A shift project involves engineering, business and behaviour changes that accomplish the down-shift in energy and resource use. The shift projects are also aimed at improving quality of life, environmental conditions and other factors, so if a monetary value can be put on these factors, they can be included. Generally, however, to keep the decision support analysis simple, the financial analysis will focus on operating costs that the organization currently incurs, particularly from energy use, but also from water, infrastructure costs and waste disposal. Once the shift project concepts have been explored and the final design chosen, the energy and resource step-down is calculated using standard energy engineering analysis. Once the energy, resource and waste step-down is quantified, it is assumed that this step-down will persist for the service life of the improvements and that the maintenance and replacement will preserve the step-down. In other words, once you make a step-down, you would not go back to carrying out the same activity and use more energy and resources to do the same operations.

7.2.1 InTIME Financial Analysis Method

The idea at the heart of the InTIME financial analysis is to provide decision-makers with two perspectives at different times in the future and ask, 'Which position is better?' The analysis considers if the decision to carry out the shift project in the current year was good from the perspective of the business in the future. We don't discount the future; we visit the future and evaluate the current alternatives from that future perspective.

Consider the example of the commercial scale solar PV plant development shown in Figure 6.4 in Chapter 6. The 4-MW capacity system in Reno, Nevada, has a utilization factor (U) = 18.9% and CU = 100% and produces 6,609 MWh/yr. The conventional analysis for 15 years gives a large negative net present value (NPV) = −$8.1 million. If we assumed a 30-year life and replacement of inverters every seven years, the NPV is still negative: −$7.05 million. For the 30-year InTIME analysis, first let us set up equal annual payments of $600,000 to finance the initial PV cost and inverters. Next calculate the net revenues in future years (electricity sales at $0.097/kWh, with 2% inflation minus $600,000). The results are typical. NPV gives much more weight to the initial cost and heavily discounts future benefits and costs. Adding up all the profits gives a total over 30 years of $4.7 million, which is a much different view of the project than the NPV. In year 15, we would make the finance payment and collect revenues, and we would have $125,703 in profit. If we didn't build the system, we would have no profit. The profit from solar PV electricity sales after paying the finance payments is as follows:

Year 5	$29,959
Year 10	$55,406
Year 15	$125,703
Year 20	$203,316
Year 25	$289,008
Year 30	$385,618

The InTIME financial analysis looks at two cases – either the shift project is carried out and the step-down is realized, or the shift project is delayed. Table 7.2 describes the steps of the InTIME financial analysis. The analysis looks at the relative competitive position of the transition organization that undertakes the shift now, or the BAU organization that undertakes the shift in the future. The perspective will be shifted into the future in 5 years, 10 years, 15 years, 20 years and longer, depending on the nature of the organization and the shift project. In each of these future years, the pressures of FOE should be applied to assess the value of the shift project in mitigating risks of price rises and resource availability problems in the organization's operations and in its supply chains. Recall that FOE has increasing probability of supply issues of greater impact with increasing time.

TABLE 7.2
InTIME Financial Analysis for Shift Project Options – Competitive Position Forecasting

Time	Perspective	Analysis
Present	Design Shift Project and model reduction in energy and resource demand. Conventional expectations for future. Decision is whether or not to invest in a shift project.	Calculate costs of the shift project and savings NPV for shift and for BAU Payback period for shift project Internal rate of return for shift project Assume 10-year financing of the shift project
5 years	There are now two organizations. The BAU organization did not take any action 5 years ago. The transition organization made the shift. Apply a 50% energy price spike for year 5.	Set up a spreadsheet with the annual expenditures for the BAU and transition companies. Explore different energy inflation rates. Compare the cash flows in this year between the two organizations.
10 years	The two organizations are still carrying out the same operations. The transition organization is now finished with the financing payments. Looking back at the last 10 years, which organization is in a better position?	Calculate the input costs per unit of good or service for both entities. Clearly, the energy costs for the transition organization are now much lower than those for the BAU organization. Evaluate the competitive positions if demand for their products declines by 20%.
15 years	The two organizations are carrying out the same operations. The BAU organization now is considering undergoing the same shift project to reduce energy costs.	What is the difference in production costs between the companies? Apply a production or throughput increase of 10%. Which organization is in a better position to increase revenues?
20 years	Considering that the organizations are doing the same work, what percentage more per unit of production does the BAU company have to earn in revenues compared to the transition company?	Assess the probability that the BAU company can compete with the transition company for the same customers. Calculate the unit production cost if the demand reduces by 25%. Is the transition company more resilient to recession and to a decline in demand?

7.2.2 Competitive Position Forecasting (CPF)

Transition economics provides the long-term view of alternatives to decision-makers from the perspective of different points in the future. Comparative competitive forecasting is aimed at management of the future position of the organization. Business competitiveness related to undiscounted stocks and flows, revenues and expenditures is analyzed in the future, when they take place. The decision to carry out an energy transition now results in future commitments to energy demand or savings, which can affect future profitability and competitive position compared to a business that made the contrary decision. The business position in 5, 10 and 15 years for a given transition management decision is compared to the positions of a hypothetical competitor who made the alternative decision, as described in Table 7.2.

Figure 7.7 shows the annual expenditures for two companies with similar operations, making the same products with the same annual production volume of 200,000 units. The transition company carries out a successful InTIME shift project and invests $100,000 to change its operations. It down-shifts energy, water and materials use and waste disposal, reducing expenditures from $16,000 to $6,000 per year. The BAU company carries on with business as usual. The cash flow for each of the ensuing 20 years with a discount rate of 10% and an inflation rate of 2% gives NPV(Transition) = $158,434 and NPV(BAU) = $155,825, which are nearly equal. The simple payback period (PP) = 10 years for the transition project investment.

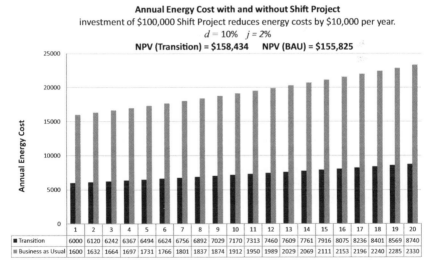

Annual Energy Cost with and without Shift Project
investment of $100,000 Shift Project reduces energy costs by $10,000 per year.
$d = 10\%$ $j = 2\%$
NPV (Transition) = $158,434   NPV (BAU) = $155,825

	1	2	3	4	5	6	7	8	9	10	11	12	13	14	15	16	17	18	19	20
Transition	6000	6120	6242	6367	6494	6624	6756	6892	7029	7170	7313	7460	7609	7761	7916	8075	8236	8401	8569	8740
Business as Usual	1600	1632	1664	1697	1731	1766	1801	1837	1874	1912	1950	1989	2029	2069	2111	2153	2196	2240	2285	2330

FIGURE 7.7 Energy expenditures in future years in current year dollars, subject to 2% inflation, for two similar companies. The transition company makes a capital investment to reduce energy, while the BAU company does not. The NPV of the capital and future costs at a 10% discount rate is nearly the same for the two companies.

The internal rate of return (IRR) = 3.6%. Thus, conventional economics would suggest that there is no great reason to do the transition project. After all, the shift project has not changed the product or the throughput, has roughly the same NPV as doing nothing, and doesn't pay back for 10 years.

The following arguments for forgoing the transition project would commonly be given:

1. The shift project does not have sufficiently short payback period. Payback actually means that the shift project is *free*. Does it matter if it is free after 2 years or after 10? What is the payback period of *not* doing the shift project? Clearly having a payback is preferable?
2. The shift project does not have a sufficiently high rate of return. The shift project has a return. There is no return on BAU, only expense.
3. NPV of the production costs for the transition is no better than BAU. The discount rate used to calculate NPV reduces the value of future benefits and costs and gives full value to initial capital costs. The underlying assumption of the NPV analysis is that the future doesn't matter. It is a mathematical construct to disregard the perspective that the company will have in the future. Even with inflation in energy costs, as used in this example, the large differences in annual costs between the BAU and transition companies is not evident in the NPV calculation, even though the disparity gets larger with time.

This example illustrates the dilemma that energy management engineers and sustainability champions have faced for more than 30 years. Projects to improve efficiency are well known. Energy-efficient appliances, lights, windows and so on are available, and the energy savings can be modelled. However, decision-makers all too often elect not to undertake the project because the rate of return is too low (the hurdle rate is usually 8%) and the payback period is too long (more than 2 years is not considered attractive).

The InTIME shift project requires a large capital investment. Let's assume that the $100,000 was financed over a 10-year term with a 0% interest clean development loan. The transition company would have the same operating costs as the BAU company for the first 10 years. In year 10, the capital expense would have been repaid, so at that point, the transition company would have costs of production 63% lower than the BAU company; by year 15, that advantage in production costs would have increased considerably. Let's say that, in year 5, there is a spike in energy costs of 50%. The transition company would then have production costs of $10,000 + $6,494 × 1.5 = $19,741, while the BAU company would have production costs of $17,318 × 1.5 = $25,977. The shift project has provided *mitigation of risks* of resource and waste management cost increases that could protect the transition company's profitability.

At the 15-year perspective, we look at the cash flows for both companies and ask if the transition company is in a better position. In year 1, even with the financing expenses, the transition company had lower expenses, by $200, than the BAU company. At the 15-year point, the BAU company is spending more than $13,000 more than the transition company to do the same business. The BAU company spent more on the inputs every year for the first 10 years, even though the transition company was paying off the capital investment. Adding all of the additional expense by the BAU company over the 20 years, the total is $142,974, almost 50% more than the capital investment in the shift that the BAU company did not think was worth it.

At this point, the two companies are making the same product in the same volume, a product that sells for the same price. At the 20-year point, both companies are producing 200,000 units and generating the same revenues. Assume that each unit sells for $1 and the labour, rents, sales and other operating costs add up to $0.50 per unit for both companies. The profit margin for the transition company is (profits/revenues) 45.6%; for the BAU company, it is 38.3%. If the recessionary pressures of the FOE cause demand reduction of 25%, then the transition company profit margin reduces slightly, to 44.2%, but the BAU company is hit harder and its profit margin drops to 34.5%. If you were the owner of the company at that time, which decision 20 years ago looks like a good idea? Essentially, with the end of cheap energy and water and waste disposal, and with the increasing probability that demand will decline into the future, companies must practice transition management and undertake energy shift projects, or they will lose out to those companies that do. This will be one of the main arguments that you will learn to develop and articulate in the InTIME financial analysis.

7.2.3   FINANCIAL OPPORTUNITIES FOR SHIFT PROJECTS

Most professional energy engineers have experience with decision-makers electing not to carry out energy-efficiency measures due to a 'long' payback period. This decision means that they are actually committing to high costs forever rather than a high capital cost short term, recovery of that investment cost and reduced costs forever. The future blindness induced by conventional economics means that companies are willing to pay higher prices over a long time period rather than invest in change. This actually presents an opportunity for a new kind of business that invests in and generates profits from energy shift projects.

The InTIME methodology discovers an opportunity to make the energy shift and reduce fossil fuel consumption by 80%. The Transition Engineering business value proposition to the customers is a hedge against inflation in energy price. The transition engineer will be paid the current rate for energy and in-turn finance and carry out the shift project and pay the energy bills. Thus, the profit for the transition goes to the Transition Engineering company that was willing to take the risk of looking into the future and see that it is, in fact, not discounted. Many of the wicked problems of the energy transition require innovations in business in order to carry out the shift project because the BAU business does not recognize the opportunity.

First, recognize that there are investment funds looking for socially responsible options; carbon credits are purchased by emitters that can be invested in your shift project if you can document the energy shift achieved. Standard methods and calculation tools are available from many sources to calculate the carbon reduction from reduced electricity, fuel and material use in your area. Second, energy management is the main service that the shift project offers. The opportunity is to offer value to the customer of a known or 'hedged' forward energy cost contract. The InTIME contract will be to design and carry out the energy shift project. The contract will have a long-term local energy network operation and management agreement. Ideally, the InTIME service will bundle a group of houses or businesses together, and manage them like a sub-network. Capital investments like high-performance windows will add value to the property, so the InTIME service will also negotiate a lien on the property for an agreed amount related to the capital improvements. The InTIME service will broker a power purchasing agreement with the local utility and negotiate a favourable rate structure, including time of use and peak demand management. The revenues for the InTIME service arise from the difference between the energy costs from the utility and the fixed contract payment from the client company.

Let's return to the example of the BAU company producing 200,000 units of goods and spending $16,000 on energy inputs. The InTIME service invests $100,000 and gains a contract for 20 years to supply the energy to the BAU company at a rate of $16,000 per year with a fixed inflation rate of 1%. The value to the BAU company is that it is insulated from price volatility in the energy inputs, it can plan on a known price, and it gets new equipment that it agrees provides ancillary benefits to it and its workers. The income stream for the InTIME service is the difference between the BAU and the transition case for the next 20 years. The total earnings for the InTIME service on the $100,000 investment is $209,261, which would be an annual rate of return of 5.5%. If the fixed inflation rate were 2%, then the earnings rate would be 6.5%.

7.2.4 INDUSTRY LEADERSHIP

There are numerous examples of large companies investing in transition management. The book *The Big Pivot* (Winston 2014) explains how companies are addressing the mega-issues of climate change and resource-carrying capacity. Transition economics recognizes the value created in reducting consumption in the supply chain, the commercial and manufacturing operations, *and* in consumption of products and resources by consumers.

Table 7.3 gives the high-level goals of some major global companies. Most of the stated goals or targets are quite sound, and many are aligned with the COP21 agreement requirement for emissions reduction. However, it is clear that companies are still struggling with *how* to achieve their goals. The InTIME method does not propose any particular green energy technology as a solution for energy transition any more than the Army Corps of Engineers proposes that concrete dams are the answer to all geotechnical problems. Transition engineers do not claim to have simple solutions; rather, they offer a means for discovering changes that meet environmental goals and improve value. The Global Association for Transition Engineering (GATE) provides training to engineers, and shared knowledge and professional support for transition engineers.

TABLE 7.3

Examples of Major Corporation Targets, Goals and Accomplishments in Transition Economics

Company	Country	Goals or Target Statements
British Telecom, 2007	United Kingdom	Cut carbon emissions by 80% by 2016. Net benefit: help customers reduce carbon footprint three times more than company footprint reductions. Savings from efficiency improvements $33 million in 2012 alone.
Ford Motor Company, 2005	United States	Double fuel efficiency by 2025. Strategy: lead market with hybrid and electric vehicles, smaller engines, lighter vehicles. Develop products ahead of competition and regulations. Statement that the need for personal vehicles must be reduced.
Interface Carpets	United States	Carbon-free product manufacture and restorative product lifecycle.
Hewlett-Packard	United States	Net positive: value-chain goals reduce company and customer footprint.
Dell Computers, 2013	United States	'[B]y 2020 the good that will come from our technology will be 10 times greater than what it takes to create and use it'.
Disney	United States	Target – Net positive impact on biodiversity and ecosystems.
LG Electronics	Korea	50% cut in GHGs
Mars	United States	Use no fossil fuels and emit no GHGs by 2040.
Apple, P&G, Walmart	United States	Goals set for 100% renewable electricity, no target date set.
IKEA	Sweden	100% renewable energy by 2020.
Nestle	Switzerland	100% renewable energy.
BMW	Germany	100% renewable energy, no target date.
Sony	Japan	Road to zero plan: zero impact across the company's value chain by 2050.
EMC, 2005	United States	Goal: 80% carbon reduction by 2050 with absolute peak in emissions by 2015. Energy intensity per revenue reduced 40% from 2005 to 2015.
Toshiba	Japan	Improve eco-efficiency (GHGs, resources, chemical substances) tenfold between 2000 and 2050.
Diageo NA, 2008	United States	Target: Cut emissions 50% by 2012. Accomplishment: 80% emission reduction as of 2014.
LEGO	Denmark	100% renewable energy.
Unilever	United Kingdom	Sustainable living plan: meet UN requirements to reduce GHGs by 50%–85% by 2050.

Note: Most of the GHG emissions reductions reported were through energy efficiency improvements.

7.2.5 MULTI-CRITERIA DECISION ANALYSIS (MCDA)

Decision-making is the process of observing and gathering information about options, analyzing the context and motivations around the decision, and applying a strategy to reach the decision. For example, a simple decision about what to wear to work requires observing the clean clothes in your wardrobe and consulting your calendar to see what you need to do today. The decision may be influenced by the objectives you have about the future; for instance, you may be interviewing for a position. You also need to consider the other clothes you have, what your schedule is for the days until laundry is done, and which clothes would be most appropriate for the FOE. There is also uncertainty to deal with, for example, the weather, which depends on analysis of scientists making predictions based on weather models. Hypothetically, if you had perfect knowledge about all of the factors, your decision would still involve your best guess about how the choice of clothes could affect the outcome of the interview, or the chance that it could rain.

The strategic analysis method described in the previous chapter is one kind of MCDA. Engineers will be familiar with the Pugh matrix used in design to evaluate design concepts and choose between a list of possible features. The evaluation process needs to have criteria that form a baseline for the desired outcome or performance. The general arrangement for a Pugh matrix is where the alternatives (A, B, C, D) for achieving the same outcome are arranged on the top row, and the various criteria (1, 2, 3, 4, 5) that are important in achieving the outcome are arranged in the first column. Each of the criteria may be weighted with a factor reflecting its importance. The baseline for the outcome is set as zero, and how each alternative meets each criterion is assessed and assigned a simple value of meeting the baseline = 0, being much better than the baseline = +2 or being much worse = −2. The importance factor times the assessment rating of each criterion for each alternative then generates the Pugh matrix. The Pugh matrix is then used for team discussions and explaining the reasons for recommendations to decision-makers. It is used as a discussion tool, not a computational model.

Risk attitude, benefit-to-cost ratios, and lifecycle cost evaluations are common quantitative assessment tools to use in decision analysis. However, decision-making depends to a large degree on the baseline and objectives, and also on the perspective and past experience of the decision-makers. We see wide cultural variations in the way people make decisions. Our interest in this text is the transition to sustainability. It is good to consider here that there have been many civilizations that have achieved a way of life that, by any objective measures, would be considered sustainable. The decision support tool used by sustainable societies is tradition. Traditions are used to govern, reinforce and teach the way that things work. Traditions are hard to change, and they are a good way to manage risk. A notable common factor of sustainability in world history is the absence of rapid technology change. As engineers, this fact challenges us to consider our role in society. Making things work is the prime directive for engineering. Making things work can be accomplished by changing how things work rather than adding new things. One perspective on sustainable energy is production of more energy from renewable resources and building more electric vehicles and batteries. Another perspective on sustainable energy is using less energy from fossil resources and driving the existing vehicles much less. The challenge for the energy transition is to try to use all perspectives but to keep a tight focus on the baseline objectives.

7.3 DISCUSSION

During this period of transition, it may seem like carrying on with business as usual is an option. Decision-makers in most companies are of an age that their entire experience is shaped by the extraordinary growth in wealth and technology that has occurred since World War II. There are no people alive today who remember when people spent more time walking than driving. It is essential to learn the lessons of history, but the creativity needed for the energy transition must come from people who understand that winding back the clock is not an option. Transition Engineering is about going forward. But it is about going forward with the perspective of a mountaineer as opposed to a person at Disneyland. Climbing a mountain and riding the killer coaster are both entertaining, and both require a positive attitude. However, the mountain climber must be realistic and aware of the risks. The mountain climber must plan to bring what she needs, and she must be prepared with maps, the right equipment and good conditioning. The adventures at Disneyland are more like the BAU path where a tourist carries on, follows the signs, behaves per instructions, keeps arms and head in the car at all times and gives up when he is finally exhausted.

Transition Engineering is definitely about positive progress; it is also about having realistic expectations, creating new knowledge and finding the determination to take on wicked problems. Transition Engineering will not be accomplished sitting down or waiting in lines. A company, community or organization that puts its adaptive capacity and resourcefulness to work to discover and carry out energy shift projects will have mitigated many of the risks to its essential activity systems that are going to result from pressures for change over the next 100 years. Figure 7.8 illustrates the difference between the work of energy transition and carrying on with BAU. The work of Transition Engineering will not be easy.

Energy transition involves a vast array of short-term projects in existing systems to down-shift long-term costs and impacts while meeting immediate needs. Focusing on 'ten things you can do to save the planet' may actually increase the risk by rushing more sustainably towards catastrophic unsustainability. The documentary film, *The Inconvenient Truth* (2006), which featured former US vice president, Al Gore, laid out the big picture of the problems of global warming due to emissions of GHGs and what the implications of climate change could be. The film had a widespread influence as far as presenting the ultimate problem of global warming and understanding of the nature of the catastrophic failure. However, at the end of the film, 'ten things you can do to save the planet' were presented with a stirring musical theme song.

- Switch to green power: use solar, wind, geothermal and biomass
- Vote with your dollars: use brands and stores that are making efforts to reduce their emissions.
- Support an environmental group.

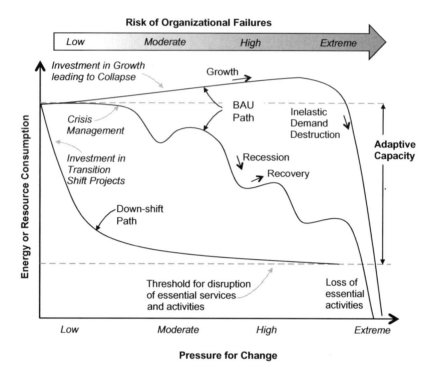

FIGURE 7.8 The decision to carry out the energy transition through engineered shift projects is an opportunity to design for down-shift to preserve essential needs. The transition down-shift path has lower risk than the inevitable decline of the BAU path through successive crises or instances of demand destruction.

- Purchase offsets to neutralize your remaining emissions.
- Buy things that last.
- Modify your diet to include less meat.
- Don't waste paper.
- Carry a refillable water bottle.
- Telecommute from home: reduce the number of miles you drive.
- Buy products that use recycled packaging or that have reduced packaging.

Of course, all of these and other actions, like improving energy efficiency by purchasing new light bulbs, could and should be done. But the use of fossil fuel is causing accumulation of GHGs in the atmosphere. Whether the fossil fuel was used to make a wind turbine or electric car, it accumulates in the atmosphere just the same. Let us be clear, the project of energy transition is to down-shift all consumption in high-consuming societies, in particular the United States and other western countries. We will end this chapter with a thought experiment about low-hanging fruit.

7.3.1   LOW-HANGING FRUIT

You have an apple tree with ripe fruit. If you want to eat an apple, you can just reach up and get one. An apple has 45 kcal and you expend 1 kcal to raise your arm, so your EROI is fantastic. You can reach a lot of apples without even getting out of your chair. As you pick more apples, the economy of scale makes apples cheaper, so you just take one bite, throw it on the ground and pick another.

Your neighbour comes by and tells you about the farmer's market in the village. He says you should get a basket and take apples to the market and make some cash. You have to get out of your chair, buy a basket, and spend 1 hour picking and 5 hours at the market. But you make enough to pay back the basket on the first weekend. And after that you are out of debt and earning cash. You buy some cheese, and use the basket to carry the cheese back home from the market. You continue picking a basket-full each week and taking apples to the market. You buy a hat and a lawn ornament and enjoy your prosperity.

Then a baker from the village asks you to double your apple production because she is making apple pies and they are selling really well. She agrees to pay you directly for the extra basket of apples so you don't have to take them to the market. Now you have a wholesale customer too. This saves you time at the market, and without renting a bigger retail stall, you can sell more apples. So you buy another basket, and you have to hire a labourer to help you pick and a wagon to carry the apples to the village. Your profits go up a bit, and now the village economy is growing. People are even coming into the village on market day from other places to have a pie and do some shopping.

After several weeks, however, the apples on the lowest branches of your tree are gone. You will have to invest in a ladder in order to pick any more apples, and it will take you and the labourer longer to pick them. You can see that there are about as many apples in the middle branches of the tree as on the bottom branches, so you can probably pay off the ladder, and you will definitely need the labourer to hold the ladder in the riskier higher-hanging fruit operation.

You wonder if it is worth it. You won't make as much profit on those higher apples, you have to take the risk of climbing a ladder into a tree, and you will have to work harder. You call the baker and tell her that you are probably done selling apples. She gets quite upset! She has bought a new pie oven in order to meet the demand for pies, and if you stop selling apples, how will she pay it off? She offers to pay more for your apples. You decide that you can afford the debt for the ladder if you can sell the apples for a higher price. Now you and the baker are in debt, but you buy the ladder and carry on. Your apple production is the highest it has ever been, but it is not expanding anymore. You have reached peak apple picking.

After a few more weeks, you notice that there is another apple stand at the market and people are going to that stall as it is offering a lower price. You have to lower your price or risk not selling enough to pay your debt on your ladder, and you can't afford any cheese. The next week you notice that there are new tables and umbrellas at the market that the mayor has installed due to the growth of the market. Now the whole town is sharing in the prosperity because you brought resources to the market. You feel important when the mayor comes to see you and tells you about the second baker who has started making pies and the expansion of the farmer's market.

A couple of weeks later, you bring your last baskets of apples to the market. You have picked all that you can reach with your ladder, and you have almost paid off the ladder. You let the market manager know that you won't be needing the stall anymore, and you post an add to sell a second-hand ladder. Suddenly the mayor, the bank manager and the bakers are at your stall, and they are quite upset. The whole economy depends on apples. There has been another business just built that is pressing apples to make cider, and they have taken out a big loan from the bank for the equipment. The city has issued bonds to pay for a new parking lot for all the new traffic at the market, and the inn has added new rooms to accommodate the new visitors to the town. There are even two people who have converted their farms from a mixture of dairy, vegetables, poultry and grains into vast apple orchards, but it will be a few years before they are in production. The village can't lose its market position now! The village has made it into the travel magazine as the apple pie experience everyone must have; you must be able to supply more apples.

You are a bit perplexed. Surely it was obvious to everyone that apples are a finite resource. But you decide to pick the few apples remaining on the top branches. You would have to get a cherry picker crane to raise you up to the top, but then it would be dangerous to reach the apples. The mayor promises funding to any inventor who can develop an apple picking arm extender so you can reach the apples at the top. The bank offers a loan for the cherry picker machine. But even with this huge investment and new technology, which isn't even working yet, you could only get a few apples because *that is all there is*. The last few apples at the top will cost so much more to produce than the ones at the bottom. What price are people willing to pay at the market for the last apples to make it worth buying a cherry picker? Is the cherry picker useful for other things or will it become a *stranded asset* like the ladder? When is it not worth it to invest in enhanced recovery to continue the diminishing extraction of a finite resource?

Unfortunately, the prospect of peak and decline of apple supply is not going over well with the businesses, council or mayor. The high-technology solution is not working out, even as the pressure to produce apples grows. Should you do what the market requires, even if it destroys the future of the resource? When does it make sense to go into debt to maintain production? When does it make sense to destroy the long-term viability for short-term gain?

In this story, we have used apples to understand resource development, growth of markets, the economic marketplace, the nature of debt, and the unsustainability of growth of demand. We have also seen how EROI works. We want a solution to the wicked problem we have all created by doing good things. Can't we just get another tree? Isn't better technology the answer? Can't the scientists invent artificial apples? We could justify taking on debt now to buy the cherry picker because there will be more apples to harvest in the future. But you don't actually need the cherry picker for the low hanging fruit in future years either.

Your prosperity as an apple producer increased when you invested in a basket. Your revenues increased when you invested in another basked, a labourer and a wagon. Jobs were created and the village economy also grew. But when you bought a ladder, you were investing more just to keep the same operation going. This is where the peak of production had been reached, but also the peak of the economic prosperity. If you had bought the cherry picker, you would have a large debt but fewer apples to bring to the market. The investment in the new technology to try to reach the last of the fruit will not prolong prosperity; it will bankrupt you, and the local economy will not materially benefit either. Of course, the most unsustainable action would be to purchase the chainsaw, which is affordable, and get the last few apples that way.

At first, when you produced more apples you were more prosperous. When the apple supply was large, the price at the market was low. Then, as the supply declined, the price went up. But your costs of production went up faster. You definitely stopped wasting apples when you stopped taking one bite and throwing them on the ground. Your increased productivity brought a few more apples to the market. Maybe future productivity gains could be realized by cutting the labourer's wages. At what point does the prosperity of the whole economy (you, the baker, and the pie-eating community) actually get better if you *don't* invest in the cherry picker or the chainsaw? The apple supply is going to run out anyway, the labourer will be out of a job and the baker will have to stop making apple pies. You might as well get the last bit of profit. You could keep the apple economy going one more week and worry about the future later.

Let's change the ending to the story about low-hanging fruit. In transition economics, we think about time as a continuum of value rather than as a boom and bust. In the story, transition engineers would look for balance of the debt and the benefit. The FOE of apple production is obvious. The investment in developing and purchasing new technology to get the last apple is not warranted. Transition engineers would also advise the mayor to limit any further conversions of farms to industrial apple production. The current number will supply the cider business, but any more conversions would reduce the local food supply, the farmer's market would struggle and people would have to start shipping in food from elsewhere. The first shift project is a baking festival with games and a dance competition. People can pay the baker to use the oven to bake the diverse range of pies they bring for the competition. The market keeps going with new people coming to shop, watch the competitions and taste all the variety of pies. And you can go back and sit on your chair until next year when it is apple season.

Is it worth it?

8 Conclusion and Discussion

Learn from mistakes.

I have read a good number of books about climate change, peak oil, economics post-growth, social change and environmental crisis. The authors usually have good descriptions of the problems; the science; and the issues with politics, consumers and economic theories. And then they present their last chapter. These authors always want to give the answer at the end after explaining the risks and critical nature of the problems. I have been greatly disappointed and even frustrated by the last chapters in these books. The authors present 'things we can do', but they are usually not technically possible or advisable. Solar PV, wind, nuclear, biofuel, hydrogen, Internet of Things, carbon capture and storage (CCS), carbon tax, electric vehicles (EVs) and distributed generation and storage are the usual great answers put forward to give us hope. In your study of Transition Engineering, presumably you understand what we can and can't do with these technical solutions. Presumably you have accepted the initial premise I put forward in the preface to my own book that engineers in all sectors are responsible for changing the existing systems they are knowledgeable about in ways they know will work, with down-shifted energy and material use. I hope you have come to the realization that we will have to do this without direction from economists, politicians, business leaders or the public, as I illustrated with the *Titanic* story in the preface.

Chapter 4 presented the seven-step method for unlocking your ingenuity. This method has worked every time we have used it. We have discovered unexpected and disruptive ideas that can be implemented in real, viable and rational shift projects. In every instance so far, the shift project has clearly been in the direction of a down-shift in material and energy consumption, and produced increased benefits, regeneration and real value creation. Here in the last chapter of *Transition Engineering: Building a Sustainable Future*, I am presenting *the answer* like all other books. But my answer isn't something that someone else should or must do. The answer is that the engineering and applied science disciplines will evolve. Just like many instances in the past, but most notably in the case of Safety Engineering 100 years ago, the change of perspective and practice in the engineering fields will change.

In Chapter 5, I put forward the theory of anthropogenic system dynamics, which holds that societies, economies and day-to-day behaviours adapt quickly to technology changes, but they can only adapt; they cannot cause the change. The people who change the built environment and technology are you and me, the engineers, technologists and fixers. We can definitely respond to social imperatives, and that is what we will do in the transition period. We must work within the political and business context, but we will also change that context, as explained in Chapters 6 and 7.

We will discover, design and build the down-shift in every energy and material system. That sounds great and hopeful, and it also sounds like job creation, R&D projects, new education programs and lots of important work. Science, technology, engineering and mathematics (STEM) geeks can save the planet!

We have plenty of evidence that planet Earth wants saving. National and local governments around the world are setting aggressive targets for emissions reduction. Cities are declaring climate emergencies. Companies are setting emissions targets and hiring consultants to provide sustainability indicators. Climate action groups are calling for 100% renewable energy. Societies around the world are saying loud and clear that they want a future under 1.5°C, they want polar bears, and they want sustainability. STEM geeks have the permission of the passengers to save the ship!

But how do we do it? Aren't all of the forces of ignorance, greed, politics, self-interest and the inertia of business as usual (BAU) aligned against us? Don't people really want something they can't have – the illusion of green growth by substitution of clean energy magic, EVs and biodegradable disposable plastics? Who is going to pay us to stop shovelling coal into the boiler and slow down the ship?

The way forward is the same as it was 100 years ago. The way that STEM professionals do their work will change because the professions will evolve. The evolution of the profession will progress in the ways they always have before:

- *Research*: Researchers recognize the knowledge gaps and the need for new types of data, analysis, modelling and design tools. The next generation of research students will develop what the industry needs to do the job. Journals start to serve the field, and the experts start to make names for themselves. Universities establish research centres, and governments and companies put funding into the Transition Engineering research projects. Leaders start a network of universities and national labs. Conferences bring people together to share ideas, debate and challenge each other – but they use a new virtual conference platform so participants do not have to fly internationally. Within a few years, the new PhD graduates are in high demand for teaching and research at major universities and institutes.
- *Education*: STEM students at universities learn about the new perspective and learn the seven-step method, along with all of their standard math, science and engineering coursework. Textbooks are written, and more and more examples are generated. Professional development and continuing education courses are developed and spread quickly. Companies send employees to the interdisciplinary transition innovation, engineering and management (InTIME) programs to discover disruptive shift project ideas for their companies. Philanthropists offer huge numbers of scholarships to attract students to Transition Engineering Programs.
- *Profession*: Consulting firms offer Transition Engineering services. A new professional organization helps to define the InTIME working standards and practices. The demand for InTIME work grows as companies seek to transition their products, operations and businesses. Professionals from the range of disciplines get together in transition courses and workshops, where coaching and group collaboration can help them with the wicked problems in their organizations.
- *Policy*: Governments set up monitoring systems to track the progress of the down-shift towards their targets. The previous national targets are upgraded. The work of government becomes facilitation of regenerating, rebuilding, redeveloping, replanting, recovering … and developing all of the new regulations and incentives that go with these new down-shift activities.

- *Society*: People adjust. Down-shifting adds resilience to risks and supply uncertainty. People come to refer to themselves as citizens, or by the name of their productive activities rather than as 'consumers'. Job growth in primary sector family businesses is strong as industrial agriculture is disaggregated and regenerated. Mass production is reimagined in a franchised pattern through small, local, efficient provision of services rather than mass marketing. Language starts to change and people don't use words like 'the economy' when they mean their city, business, environment or country.

8.1   THE BIG DO

Transition Engineering: Building a Sustainable Future has been about how down-shifting can be discovered and carried out as shift projects. In this last chapter, I have provided a vision of how the energy transition can be accomplished by the evolution of the engineering professions to include InTIME practice. But *why* would it happen? The biggest emitters are also the most profitable businesses, and the most profitable companies have access to political and financial power. How can down-shifting possibly get done? Most people don't really want things to change, even if they understand the mega-problems.

8.1.1   THE ULTIMATE WICKED PROBLEM – OIL

Oil is possibly the touchstone wicked problem. Oil is the most useful energy resource ever discovered. But the extraction and combustion of fossil carbon is the ultimate reason for the unsustainable energy supply and global warming. Oil has been the most profitable industry in history, but it has indebted, impoverished and degraded the communities where it is extracted and refined. The oil industry has immense influence and political power. But control of oil reserves and supplies has been the root of major conflicts and wars. Cars, ships and airplanes provide unparalleled mobility and freight transport. But internal combustion engines and oil production produce harmful ground-level pollution. The world's economies run on oil, but oil and transport infrastructure costs are unsustainable. Oil consumption must be reduced, and yet it cannot be reduced without catastrophic economic impact, as was demonstrated during the Organization of Petroleum Exporting Countries (OPEC) oil embargo. Oil is indeed a wicked problem.

8.1.2   InTIME APPROACH FOR OIL

My research group of transition engineers at the Advanced Energy and Material Systems Lab (AEMSLab) in Christchurch, New Zealand, carried out a seven-step InTIME investigation of the oil industry in late 2018. One hundred years ago, the age of oil was just beginning. Extraction, refining and end-use technologies were at early stages and were inefficient and dangerous. But the development of better engines and the economic and technology boom after World War II led to exponential growth in oil production and consumption. The fields of petroleum geology and petroleum engineering emerged. The Society of Automotive Engineers (SAE) was founded in

New York City, United States, in 1904 with annual dues of $10. The SAE provided open sharing of knowledge-based engineering, set up standards, and organized meetings and conferences. By 1916, SAE had 1,800 members. Transportation engineering, traffic and road Safety Engineering, air pollution management and many other fields related to production and use of oil also emerged. The only reversals in the exponential growth of oil use were the result of the OPEC oil embargo in the 1970s and the recent global economic crisis that was precipitated by the massive spike in oil prices. In 1965, global oil production was 30 million barrels per day (mbpd), and in 1992 when the UNFCCC agreed to reduce greenhouse gas (GHG) emissions, oil production was 68 mbpd. Global oil production is currently at 100 mbpd and growing.

The BAU scenario is for oil production to continue with slow growth, possibly peaking in the 2040s. Conventional oil has already peaked and is declining, but high prices have caused investment in unconventional reserves. The technology wedge scenario is the substitution of petroleum cars with EVs. If sales of EVs grew by 60% per year from 2018, then oil consumption reduction for driving could be 2 mbpd by 2023. But can car manufacturing increase at such rates, and can it be low carbon? The efficiency wedge scenario is for improved maintenance of vehicles and reduced highway speed. A top global driving speed of 90 k could reduce oil consumption by 4 mbpd. The fuel substitute wedge scenario is biofuel and hydrogen, but neither of these reduce future oil consumption because their energy return on energy invested (EROI) < 1. The behaviour wedge scenario is a mode shift to public transport and active modes that do not use personal vehicles. If all of the 10% of trips normally within walking or cycling distance of 2 km were shifted to active mode, then a fuel reduction of 1 mbpd could be realized. The forward operating environment (FOE) for staying below 550 ppm CO_2 is a 10% per year oil demand decline to reach production and consumption of 2.5 mbpd around 2055.

Thinking 100 years into the future is an interesting exercise, which can be done with the city of Christchurch, New Zealand, as an example. Christchurch 2120 is working well without oil. The university, rugby stadium, botanical gardens and music venues are still there in 100 years. It is a city of 300,000 people located on a relatively flat coastal plain. Electric trains and trams are in use; electric bikes are common; and the organization of activities, services and manufacturing in the city is clearly designed for active access. The rail to the hinterland and the sea port brings in food, products and materials. In general, the city is clean, quiet and safe. Much of the road area has been redeveloped into tramways, public spaces, food production and green space. Urban gardening is prevalent, but people also enjoy the public spaces and sports fields.

If we backcast, we find that there are not so many differences between Christchurch 2120 and today. The city can adapt to the world beyond petrol. This gives us the idea for the trigger. We now know that the city can adapt to using almost no oil. We also know there is no reason why it would. Thus, we will innovate the trigger, which could be, for example, that the oil companies of the world have announced that they will be reducing oil production by 10% per year until 2055.

The shift project is thus the execution of the petroleum production retreat. We used the matrix game (described in Section 5.5) to develop the shift project. The first player is the Oil Majors, representing the 12 largest oil companies. The Oil Majors have the main interest of increasing wealth for their shareholders. The Oil Majors are affiliated with different countries, and the largest are Saudi Aramco, Sinopec, China National, PetroChina, Exxon Mobile, Royal Dutch Shell, Kuwait Petrol, British Petroleum and Total. The next player is the International Energy Agency (IEA), representing 29 member countries. The IEA's purpose is to ensure cooperation and forward planning to avoid another crippling oil supply shock. The next player is the Organization for Economic Cooperation and Development (OECD), representing 36 member countries. The OECD's purpose is to ensure economic growth and fair competition. The final player is the United Nations Framework Convention on Climate Change (UNFCCC), representing 197 member countries. The UNFCCC objectives are to form agreements between countries for the equitable and sustainable development of all people. The recent UN COP21 agreement set a target of limiting GHG emissions to keep global warming below 1.5°C this century.

What are the conditions of retreat according to the players' published reports, position papers and leader statements?

The Oil Majors would retreat if they made more money in retreating than in continuing BAU.

The IEA would manage the production retreat if it was known well in advance and if the prices were set so that they could not rise or fall sharply as this causes economic disruption. The IEA would work to police the production retreat and suppress the development of a black market.

The OECD would participate in the production retreat to develop coherent policies and international agreements for allocation of the oil supply if the production rates were known at least 12 months in advance and if a stable price for oil were ensured. The OECD sees a net benefit to the global economy for this situation as it spurs innovation and investment in new technologies and developments that permanently reduce oil use, and thus stimulates activity even as the risks of future climate change costs are reduced.

The UNFCCC would sanction, negotiate and oversee the oil production retreat. Again, the conditions on forming a coalition would be to set agreed production quotas at least 12 months in advance, equitable distribution to all countries, and a stable oil price with control of black-market entrants. The UNFCCC would hold the first Coalition of Oil Producers meeting in Paris in 2019 (CoOP1) to establish the conditions for the production retreat initiation in 2020.

The shift project is this: Form an international transition collaboration to work with the Oil Majors to define the parameters of the oil production retreat. Discover and define the pathways for how they can make more money in retreat than they did pursuing growth. On first analysis, the oil industry is past the peak in investment and benefit. The low hanging fruit has been taken and now extraordinary investments are being made with ever higher risk, and lower EROI ventures are being planned. The shift project is accomplished by a successful proposal to IEA and the Oil Majors to fund a cooperative and participatory program within the transition collaboration to assess the profitability of their recent investments in exploration and drilling new reserves. The project would design the retreat and determine the parameters for increasing profitability. The project would also need to develop the algorithm for determining the stable price needed for retreat. The International Panel on Transition Engineering (ITPE) would be set up with funding from the OECD to work with member countries on modelling and optimizing policies for accomplishing the most economically efficient investments in infrastructure and business development that have the highest oil retreat index (ORI). The ORI is high for policies that have high reduction in oil demand; facilitated adaptation; and high returns on social, environmental or economic benefit. The ITPE work is carried out urgently and swiftly so that the most beneficial policies, most profitable Oil Major retreat strategies, and the stable oil price algorithm and production management mechanisms are ready for ratification at the CoOP1 in 2020.

The energy transition is possible, and it could be done, for the reasons that things are always done – profit. All those things that 'we could do to save the planet' would start to gain traction as soon as the oil production retreat was announced. Engineers with the ability to discover, design and manage the disruptive down-shift for companies and organizations would become highly popular in the industry. The membership of the Global Association for Transition Engineering (GATE) would grow, and the number of training programs and university courses would explode. The perception that the world was transitioning to a new era, through hard work, innovation, and engineering, could bring about a reversal of the recent retreat of democracy and civility in the face of climate and energy stress.

If you have read through the chapters of this book, and if you have learned how to shift your perspective from BAU to the transition trajectory, and if you have understood that there is a way to innovate, engineer and manage that transition, then you are qualified to apply for membership in GATE and start applying the InTIME method in your day job. You can be a leader, set up a local GATE and hold meetings to share ideas and learn from other members. At the time of writing in mid 2019, the most up-to-date science provided unambiguous evidence that GHG emissions are increasing, global warming is accelerating, climate changes are more extreme, and the outlook for the BAU future is more catastrophic than any previous scientists' warnings.

This is the one time in *Transition Engineering: Building a Sustainable Future* when I will use the words *we must* that I promised in the preface I would not use: We must radically change the way we, as applied scientists and engineers, do our jobs. We must take responsibility for reengineering, rebuilding, and regenerating all engineered systems and accomplish the down-shift of destruction and an increase in the quality of life, biodiversity, resilience and real value. In this case it is possible that 'we' – the engineering and technology disciplines – actually can do the things we must.

8.2 THE FINAL STORY: CASSANDRA

Cassandra was the beautiful daughter of Priam, the king of Troy. She was so beautiful that the god Apollo wanted her for himself. In order to persuade her, he gave her the gift of foresight. Cassandra got the power to see clearly what would happen in the future. But Cassandra wasn't really all that impressed with Apollo and she spurned him. Apollo wasn't pleased and decided to leave her with the gift of prophesy, but he also cursed her with talking in riddles when she tried to tell people about the future. Thus, Cassandra would know the future, but she could not deliver her warnings.

The Greeks, led by the mighty Agamemnon and the warrior Achilles, appeared to have given up their siege of Troy and sailed away. They left behind a tribute to the Trojan victors in the form of a huge horse statute. Cassandra clearly saw that it was a trap. She saw that the warriors hidden inside the statue would sneak out in the night and open the city gates and that the Greek ships would return under cover of darkness, storm the city and destroy Troy. But she was just a beautiful, young princess, and she didn't have any credentials as an oracle or a military strategist, so the army would not let her into the meeting about what to do with the tribute. She found her father and tried to warn him, but when she explained that there were Greek warriors hiding in the Trojan Horse, and that they would destroy Troy, her warning just came out in strange riddles that her father could not understand. As the soldiers opened the gates and rolled the statue into the city, Cassandra became more desperate and started screaming her garbled warnings. As she got more hysterical (from the Latin *hystera* 'womb', meaning characteristic of neurotic behaviour of women), she started to annoy everyone and she was dragged away and locked in the temple. And the rest is history.

The Greek story of Cassandra tells us a few things. First, it is not a good idea to spurn a God. Second, blessings can also be curses. But most important, delivering a warning, even when you have perfect knowledge, can be a problem. The problem can be that you don't have the recognized authority and so the people you are trying to warn don't listen. The problem can be that you are using language that people can't understand. The problem can be that your warning goes against the beliefs of the people you are warning. Or the problem can be that you get upset and your warning is rejected.

The student School Strike 4 Climate is an example of people without political knowledge calling for action that they can't explain. Scientists from the International Panel on Climate Change (IPCC) work with clear evidence, but they speak in riddles of scientific jargon and probabilities and they only tell us what we must not do. Environmental activists talk about zero carbon or substituting solar PV for oil, and this type of message is nonsensical to business people and economists. The protestors and action groups accept the science, but they are perceived as troublemakers considering their preferred methods of staging protests. Thus, we understand the Cassandra problem.

Let's change the ending of the story.

Cassandra knew what *not* to do. It was very simple: don't open the gates and roll the Trojan Horse into the city and leave it there unguarded overnight. If the statue is left outside the gates, Troy will be safe, and this is what Cassandra was trying to accomplish with her warning. If the statue is wheeled in, but the Greeks are discovered while the Trojan army is surrounding the statue, then there is a good chance that the Trojans could defeat the Greeks before they open the gate to their army. If the statue is left unguarded in the night, then the probability of stopping the disastrous outcome depends on not allowing the Greeks hidden in the statue to reach the gates and open them.

Let's change the perspective.

Cassandra knows that her warning not to open the gates and bring the statue in is not being heard. The mistake she makes is continuing to deliver the same warning in the same way at increasing volume. If she tries to tell people what will happen – her message won't be heard. Let's say she learns from her mistake. She changes her strategy of warning about what will happen. Instead, she uses her creativity to design an action that the king, the priests or the army could take. The action must prevent the future defeat, but it strategically uses the beliefs and objectives of the people with power. Cassandra speaks to her father the king about the great defeat of the Greeks and how building a platform for the statue outside the city walls would show how great he is to everyone approaching the city. If this doesn't work, she talks to the high priests about making the statue a fitting tribute to the gods as well as to the king by building a great pyre under it and sending it up in flames as the sun sets – what a great spectacle for the people! If that doesn't work, she speaks to the general about the military victory and how, for the first night, the elite soldiers should have a celebration and hold an all-night vigil around the statue with fires blazing, keeping the light on the Trojan Horse throughout the night.

Now get busy preventing what is preventable.

References

Aleklett, K., C.J. Colin. 2003. The peak and decline of world oil and gas production, *Minerals & Energy*, 18: 5–20.

Allen, D.T., D.R. Shonnard. 2012. *Sustainable Engineering Concepts, Design, and Case Studies*, Upper Saddle River, NJ: Pearson Education Inc.

Amigun, B., J.K. Musango, W. Stafford. 2011. Biofuels and sustainability in Africa, *Renewable and Sustainable Energy Reviews*, 15(2): 1360–1372.

Araujo, K. 2014. The emerging field of energy transitions: Progress, challenges, and opportunities, *Energy Research & Social Science*, 1: 112–121.

ASME. 2011. *Etp:Energy-Water Nexus—Cross-Cutting Impacts and Addressing the Energy-Water Nexus: A Blueprint for Action and Policy Agenda*, New York: Alliance for Water Efficiency and American Council for an Energy-Efficient Economy.

ASSE. 2017. American Society of Safety Engineers, http://www.asse.org/ (accessed 27 February 2018).

ASSE. 2018. http://www.asse.org/about/history/ (accessed 27 February 2018).

Associated Press. 2009. U.N.: World's hungry now more than 1 billion, 19 June 2009, http://www.msnbc.msn.com/id/31449307/ (accessed 20 June 2009).

Ballard, C.W., P.S. Penner, D.A. Pilati. 1978. Net energy analysis—Handbook for combining process and input-output analysis, *Resources and Energy*, 1: 267–313.

Barney, G.O., J. Blewett, K.R. Barney. 1993. *Global 2000 Revisited: What Shall We Do?*, Alexandria, VA: The Millennium Institute.

Bernholz, P. 2003. *Monetary Regimes and Inflation: History, Economic and Political Relationships*, Cornwall, UK: MPG Books Ltd.

Blair, N., S. Krumdieck, D. Pons. 2019. Electrification in remote communities: Assessing the value of electricity using a community action research approach in Kabakaburi, Guyana, *Sustainability*, 11(9): 2566.

Bossel, U., B. Eliasson, G. Taylor. 2003. The future of the hydrogen economy: Bright or bleak? *Cogeneration and Competitive Power Journal*, 18(3): 29–70.

BP Statistics Review. 2017. www.statsreview.bp.com (accessed May 2018).

BP. 2015. https://www.bp.com/en_us/united-states/home/community/gulf-commitment.html (accessed July 2019).

Brandt, A.R., M. Dale, C.J. Barnhart. 2013. Calculating systems-scale energy efficiency and net energy returns: A bottom-up matrix-based approach, *Energy*, 62: 235–247.

Brown, L. 2010. *World on the Edge: How to Prevent Environmental and Economic Collapse*, Earth Policy Institute, http://www.earth-policy.org/books/wote/wote_data (accessed July 2017).

Brulle, R.J., J. Carmichael, J.C. Jenkins. 2012. Shifting public opinion on climate change: An empirical assessment of factors influencing concern over climate change in the U.S., 2002–2010, *Climate Change*, 114: 169–188.

Brundtland, G.H. 1987. *Our Common Future*, World Commission on Environment and Development, New York: Oxford University Press.

Bureau of Labor Statistics. 2019. Consumer price index—All urban consumers, https://www.statbureau.org/en/united-states/cpi-u (accessed February 2018).

Campbell, C., J.H. Laherrare. 1998. The end of cheap oil, *Scientific American*, 278(3): 78–83.

Carney, B., T. Feeley, A. McNemar. 2008. *Power Plant–Water R&D Program*. Morgantown, WV: DOE, National Energy Technology Laboratory, http://citeseerx.ist.psu.edu/viewdoc/summary?doi=10.1.1.151.8516 (accessed May 2019).

Chadwick, A.R., D.J. Noy. 2015. Underground CO_2 storage: Demonstrating regulatory conformance by convergence of history-matched modeled and observed CO_2 plume behavior using Sleipner time-lapse seismics, *Greenhouse Gases: Science and Technology*, 5(3): 305–322.

Chapman, I. 2014. The end of Peak Oil? Why this topic is still relevant despite recent denials, *Energy Policy*, 64: 93–101.

Cleveland, C.J. 2005. Net energy from oil and gas extraction in the United States, 1954–1997, *Energy*, 30: 769–781.

Cleveland, C.J., P. O'Connor. 2010. An assessment of the energy return on investment of oil shale, Western Resource Advocates, https://westernresourceadvocates.org/publications/assessment-of-energy-roi-of-oil-shale/ (accessed June 2019).

Cobb, J. 2015. February 2015 Dashboard, HybridCARS, https://www.hybridcars.com/december-2015-dashboard/ (accessed July 2019).

Cornell University. 2018. http://trianglefire.ilr.cornell.edu (accessed February 27 2018).

Cox, L. 2015. The surprising decline in US petroleum consumption, World Economic Forum, www.weforum.org/agenda/2015/07 (accessed April 8, 2017).

CSI, Cultural Survival Inc. 2012. Indigenous Peoples International Declaration on Self-Determination and Sustainable Development, available at https://www.culturalsurvival.org (accessed April 8, 2017).

Dale, M., S. Krumdieck, P. Bodger. 2011a. A dynamic function for energy return on investment, *Sustainability*, 3(10): 1972–1985.

Dale, M., S. Krumdieck, P. Bodger. 2011b. Net energy yield from production of conventional oil, *Energy Policy*, 39(11): 7095–7102.

Dale, M., S. Krumdieck, P. Bodger. 2012a. A biophysical approach (GEMBA) Part 1: An overview of biophysical economics, *Ecological Economics*, 73: 152–157.

Dale, M., S. Krumdieck, P. Bodger. 2012b. A biophysical approach (GEMBA) Part 2: Methodology, *Ecological Economics*, 73: 158–167.

Deffeyes, K. 2001. *Hubbert's Peak: The Impending World Oil Shortage*, Princeton, NJ: Princeton University Press.

Diamond J. 2005. *Collapse: How Societies Choose to Fail or Succeed*, New York: Penguin.

Diaz-Maurin, F., M. Giampietro. 2013. A "grammar" for assessing the performance of poser-supply systems: Comparing nuclear energy to fossil energy, *Energy*, 49: 162–177.

Du Pisani, J.A. 2006. Sustainable development—Historical roots of the concept, *Environmental Sciences*, 3(2): 83–96.

Ecofys. 2011. *The Energy Report, 100% Renewable Energy by 2050*, www.ecofys.com (accessed March 2018).

EcoVadis. 2017. *The 2017 Sustainable Procurement Barometer*, http://www2.ecovadis.com/sustainable-procurement-barometer-2017 (accessed May 2017).

Edwards, J. 2014. Oil sands pollutants in traditional foods, *Canadian Medical Association Journal*, 186(12): 1.

Erten, B., J.A. Ocampo. 2013. Super cycles of commodity prices since the Mid-nineteenth Century, *World Development*, 44: 14–30.

Europe Africa. 2011. *Monitoring Report on EU Policy Coherence for Food Security*, available at http://www.europafrica.info/en/publications/biofueling-injustice.

Flannery, T. 2005. *The Weather Makers,* New York: Grove Press.

Flannery, T. 2014. *The Great Barrier Reef and the Coal Mine that Would Kill It*, The Guardian, available at https://www.theguardian.com/environment/2014/aug/01/-sp-great-barrier-reef-and-coal-mine-could-kill-it (accessed August 2017).

Frankfurt School-UNEP Centre/BNEF. 2015. *Global Trends in Renewable Energy Investment 2015*, Frankfurt School of Finance & Management gGmbH.

Fultz, H.E. 1958. Flashing Stop Sign, US Patent Application US2920309 A.

Georgescu-Roegen, N. 1976. (1972) Energy and Economic Myths. In: *Energy and Economic Myths*, Limits to Growth Series: The Equilibrium State and Human Society, National Research Council, Commission on Natural Resources and the Committee on Mineral resources and Environment, pp. 3–36.

Goldemberg, J, T.B. Johansson (eds.). 2004 *World Energy Assessment: Energy and the Challenge of Sustainability*, New York: United Nations Development Program.

Greene, D.L., L. Roderick, J.L. Hopson. 2011, *OPEC and the Costs to the U.S. Economy of Oil Dependence: 1970–2010*, Oak Ridge, YN: Oak Ridge National Laboratory Memorandum.

Greenstone, M. 2002. The impacts of environmental regulations on industrial activity: Evidence from the 1970 and 1977 Clean Air Act Amendments and the census of manufacturers, *Journal of Political Economy*, 110(6): 1175–1219.

Gude, V.G. 2011. Energy consumption and recovery in reverse osmosis, *Desalination and Water Treatment* 36: 239–260.

Gupta, A.K., C.A.S. Hall. 2011. A review of the past and current state of EROI data, *Sustainability*, 3: 1796–1809.

Halifax House Price Index. 2017. https://www.halifax.co.uk/media-centre/house-price-index/ (accessed March 2018).

Hannon, B., D.S. Casler, T. Blasek. 1985. *Energy Intensity for the U.S. Economy-1977*, Energy Research Group, Urbana, IL: University of Illinois Press.

Hansen, J. 2009. *Storms of My Grandchildren*, New York: Bloomsbury.

Hansen, K., D. Johnson, A. Lacis, S. Lebedeff, P. Lee, D. Rind, G. Russell. 1981. Climate impact of increasing atmospheric carbon dioxide, *Science*, 213(4511): 957–213.

Heath, G., D. Sandor. 2013. Life Cycle Greenhouse Gas Emissions from Electricity Generation (Fact Sheet). Golden, CO: National Renewable Energy Laboratory.

HM Treasury. 2004. *The Orange Book Management of Risk-Principles and Concepts*, Richmond, UK: HMSO, www.gov.uk/government/uploads/system/uploads/attachment_data/file/220647/orange_book.pdf (accessed March 2018).

Hoekstra, A.Y., A.K. Chapagain, 2008. *Globalization of Water: Sharing the Planet's Freshwater Resources*, Oxford, UK: Blackwell Publishing.

Holway, J., D. Elliott, A. Trentadue. 2014. Arrested Developments, combating zombie subdivisions and other excess entitlements, Cambridge, MA: Lincoln Institute of Land Policy.

Hovatter, S. 2013. *Brainstorming: Unleashing Your Creativity to Think Outside the Box*, Data Designs Publishing, Norwalk, OH, http://www.consumeraffairs.com/news04/2006/airbags/airbags_invented.html (accessed March 2018).

Huber, B.M., M. Cornstock, D. Polk, L.L.P. Wardell . 2017. *ESG Reports and Ratings: What they are, why they matter*, Harvard Law School Forum, https://corpgov.law.harvard.edu/2017/07/27/esg-reports-and-ratings-what-they-are-why-they-matter/ (accessed May 2018).

IEA. 2014. World Energy Outlook. https://webstore.iea.org/world-energy-outlook-2014 (downloaded March 2018).

IEA. 2015. Energy and Climate Change, World Energy Outlook Special Report. www.iea.org.

IEA. 2017. OECD Library, Energy Technology RD&D Statistics, http://dx.doi.org/10.1787/enetech-data-en (accessed May 2018).

IIASA. 2009. *RCP Database*, http://www.iiasa.ac.at/web-apps/tnt/RcpDb (accessed May 2018).

Intake: National Health and Nutrition Examination Survey. 1999 and 2000. *Environmental Health Perspectives*, 112(5): 562–570.

IPCC. 2007. Fourth Assessment Report (AR4). In: *Contribution of Working Group III to the Fourth Assessment Report of the Intergovernmental Panel on Climate Change, 2007*, B. Metz, O.R. Davidson, P.R. Bosch, R. Dave, L.A. Meyer (eds.), Cambridge, UK.

IPCC. 2013. Summary for Policymakers. In: *Climate Change 2013: The Physical Science Basis*. Contribution of Working Group I to the Fifth Assessment Report of the Intergovernmental Panel on Climate Change, Stocker, T.F., D. Qin, G.-K. Plattner, M. Tignor, S.K. Allen, J. Boschung, A. Nauels, Y. Xia, V. Bex and P.M. Midgley (eds.), Cambridge, UK: Cambridge University Press.

IPCC. 2014. Summary for policymakers. In: *Climate Change 2014: Impacts, Adaptation, and Vulnerability*. Part A: Global and Sectoral Aspects. Contribution of Working Group II to the Fifth Assessment Report of the Intergovernmental Panel on Climate Change Field, C.B., V.R. Barros, D.J. Dokken, K.J. Mach, M.D. Mastrandrea, T.E. Bilir, M. Chatterjee et al., (eds.)., Cambridge, UK: Cambridge University Press.

Jackson, T. 2009. *Prosperity without Growth, Economics for a Finite Planet*, London, UK: Routledge, Taylor & Francis Group.

Kenny, J., N. Barber, S. Huston, K. Linsey, J. Lovelace, A. Maupin. 2005. Estimated Use of Water in the United States in 2005, Circular 1344, USGS. pp. 38–41.

Kheel Center. 2014. *The 1911 Triangle Factory Fire*, Cornell University, http://www.ilr.cornell.edu/index.html (accessed February 13, 2014).

King, C.W. 2017. Delusions of Grandeur in Building a Low-Carbon Future. *Earth Magazine*, 32–37.

King, C.W., C.A.S. Hall. 2011. Relating financial and energy return on investment. *Sustainability*, 3: 1810–1832.

Krumdieck, S. 2010. Peak Oil Vulnerability Assessment for Dunedin, http://www.dunedin.govt.nz/your-council/policies-plans-and-strategies/peak-oil-vulnerability-analysis-report (accessed February 2017).

Krumdieck, S. 2011. Transition Engineering of urban transportation for resilience to peak oil risks, (November 11–17, 2011, Denver, CO) Proceedings of the ASME 2011, ICEME2011-65836.

Krumdieck, S. 2013. Transition Engineering: Adaptation of complex systems for survival, *International Journal of Sustainable Development*, 16(¾): 310–321.

Krumdieck, S. 2014. Chapter 13. Transition Engineering. In: *Principles of Sustainable Energy,* 2nd ed, F. Kreith and S. Krumdieck, (eds.), Boca Raton, FL: Taylor & Francis Group, pp. 698–728.

Krumdieck, S., A. Hamm 2009. Strategic analysis methodology for energy systems with remote island case study, *Energy Policy*, 37(9): 3301–3313.

Krumdieck, S., M. Dale, S. Page. 2012. Design and implementation of a community based sustainable development action research method, *Social Business*, 2: 291–337.

Krumdieck, S., S. Page, A. Dantas. 2010. Urban form and long-term fuel supply decline: A method to investigate the peak oil risks to essential activities, *Transportation Research Part A*, 44: 306–322.

Krumdieck, S., S. Page. 2013. Retro-analysis of bio-ethanol and bio-diesel in New Zealand, *Energy Policy*, 62: 363–371.

Krumdieck, S.P. 2017. Transition engineering. In: *Energy Solutions to Combat Global Warming. Lecture Notes in Energy*, X. Zhang and I. Dincer (eds.), Cham, Switzerland: Springer, p. 33.

Kubiszewski, I. C.J. Cleveland, P.K. Endres. 2010. Meta-analysis of net energy return for wind power systems, *Renewable Energy*, 35: 218–225.

Lambert, J.G., C.A.S. Hall, S. Balogh, A. Gupta, M. Arnold. 2014. Energy, EROI and quality of life, *Energy Policy*, 64: 153–167.

Lawrence Livermore National Laboratory (LLNL). 2019. https://flowcharts.llnl.gov/commodities/energy (accessed July 2019).

Lazard, Ltd. 2015. *Lazard's Levelized Cost of Energy Analysis, Version 9.0*, Lazard, Ltd., New York, https://www.lazard.com/media/2390/lazards-levelized-cost-of-energy-analysis-90.pdf (accessed February 2019).

Lazard, Ltd. 2009. *Energy Technology Assessment*, New York: Lazard, pp. 11–13.

Linden, M.O., A.F. Kazakov, J.S. Brown, P.A. Domanski. 2014. A thermodynamic analysis of refrigerants: Possibilities and tradeoffs for low-GWP refrigerants, *International Journal of Refrigeration*, 38, 80–92.

Loorbach, D., J. Rotmans. 2010. The practice of transition management: Examples and lessons from four distinct cases, *Futures*, 42: 237–246.

Lord, W, 1955. *A Night to Remember*, New York: Henry Holt and Company, Early Bird Books.

Loupasis, S. 2002. Technical analysis of existing RES desalination schemes – RE Driven Desalination Systems, REDDES, Report, Contract # 4.1030 /Z/01-081/2001.

Luthcke, S.B., H.J. Zwally, W. Abdalati. et al. 2006. Recent Greenland ice mass loss by drainage system from satellite gravity observations, *Science*, 314(5803): 1286–1289.

Mahaffey, K., R.P. Cliffner, C. Bodurow. 2004. Blood organic mercury and dietary mercury intake: National Health and Nutrition Examination Survey, 1999 and 2000, *Environmental Health Perspectives*, 112(5): 562–570.

Massari, S., M. Ruberti. 2013. Rare earth elements as critical raw materials: Focus on international markets and future strategies, *Resources Policy*, 38: 36–43.

Mathaisel, D.F.X., J.M. Manary, N.H. Criscimagna. 2013. *Engineering for Sustainability*, Sustaining the Military Enterprise Series, Boca Raton, FL: Taylor & Francis Group.

McCormick, L.W. 2006. *A Short History of the Airbag*, Consumer Affairs.com, http://www.consumeraffairs.com/news04/20 (accessed 2 March 2011)

Meadows, D.H., D.L. Meadows, J. Randers, W.W. Behrens. 1972. *Limits to Growth*, New York: Potomic Associates.

Meadows, D., J. Randers, D. Meadows. 2004. *Limits to Growth, The 30-Year Update*, White River Junction, VT: Chelsea Green Publishing Company.

Miotti, M., G.J. Supran, E.J. Kim, and J.E. Trancik. 2016. Personal vehicles evaluated against climate change mitigation targets. *Environmental Science & Technology*, 50(20): 10795–10804.

Mohr, S.H., J. Wang, G. Ellem, J. Ward, D. Giurco. 2015. Projection of world fossil fuel by country, *Fuel*, 141: 120–135.

Morrisette, P.M. 1989. The evolution of policy responses to stratospheric ozone depletion. *Natural Resources Journal*, 29: 793–820.

Mudd, G.M., Z. Weng, S.M. Jowitt, I.D. Turnbull, T.E. Graedel. 2013. Quantifying the recoverable resources of by-product metals: The case of cobalt, *Ore Geology Reviews*, 55: 87–98.

Murphy, D.J., C.A.S. Hall, C. Cleveland. 2011. Order from Chaos: A preliminary protocol for determining EROI of fuels, *Sustainability*, 3: 1888–1907.

NABERS, https://nabers.gov.au/ (accessed May 2017).

NASA. 2015. Global Climate Change: Vital Signs of the Planet, available at http://climate.nasa.gov/evidence/ (accessed April 2017).

NOAA. 2017. Carbon Tracker on-line resources, https://www.esrl.noaa.gov/gmd/ccgg/carbontracker/ (accessed April 2017).

NRC. 2008. *Desalination: A National Perspective, National Research Council*, Washington, DC: National Academies Press.

NREL. 2004. National Renewable Energy Laboratory Solar Energy Technologies Program Report DOE/GO-102004-1847, January 2004.

OCE Australian Government Office of the Chief Economist. 2015. *Australian Energy Update*, www.industry.gov.au/oce.

Pacala, S., R. Socolow. 2004. Stabilization wedges: Solving the climate problem for the next 50 years with current technologies, *Science*, 305(5686): 968–972.

Page, S., A.G. Williamson, I.G. Mason. 2009. Carbon capture and storage: Fundamental thermodynamics and current technology, *Energy Policy*, 37(9): 3314–3324.

Palm, W.J. 2000. *Modeling, Analysis, and Control of Dynamic Systems*, New York: John Wiley & Sons.

Pearce, F. 2013. Successful Push to Restore Europe's Long-Abused Rivers, Yale Environment 360, http://e360.yale.edu/features (accessed 8 April 2017).

Pimentel, D., S. Williamson, C. Alexander, O. Gonzalez-Pagan, C. Kontak, S. Mulkey. 2008. Reducing energy inputs in the US food system, *Human Ecology*, 36: 459–471.

Princen, T., J.P. Manno, P.L. Martin (eds.). 2015. *Ending the Fossil Fuel Era*, Cambridge, MA: MIT Press.

Prior, T., D. Giurco, G. Mudd, L. Mason, J. Behrisch. 2012. Resource depletion, peak minerals and the implications for sustainable resource management, *Global Environmental Change*, 22: 577–587.

Ram, E., CEO of Ram Energy Inc., personal discussion September 2015.

Raworth, K. 2017. *Donut Economics, Seven Ways to Think like a 21st Century Economist*, London, UK: Random House Business Books.

Rendall, S., S. Page, F. Reitsma, E. van Houten, S. Krumdieck. 2011. Quantifying transport resilience: Active mode accessibility, *Journal of the Transportation Research Board*, 2242: 72–80.

RFA. 2015. Renewable Fuels Association, http://www.ethanolrfa.org/pages/statistics#A.

Rico, M., G. Benito, A.R. Salgueiro, A. Diez-Herrero, H.G. Pereira. 2008. Reported tailings dam failures, A review of the European incidents in the worldwide context, *Journal of Hazardous Materials*, 152: 846–852.

Rojey, A. 2009. *Energy & Climate: How to Achieve a Successful Energy Transition*, London, UK: Society of Chemical Industry and John Wiley & Sons.

Rotty, R.M., A.M. Perry, D.B. Reister. 1976. *Net Energy from Nuclear Power*, Oak Ridge, TN: Oak Ridge Associated Universities.

Rubin, J., P. Buchanan. 2008. *What's the Real Cause of the Global Recession?* StrategEcon, October 31, Toronto, UK: CIBC World Markets Inc.

Sander, K., G.S. Murthy. 2010. Life cycle analysis of algae biodiesel, *International Journal of Life Cycle Assessment*, 15(7): 704–714.

Scanlon, B., C. Faunt, L. Longuevergne, R. Reedy, W. Alley, V. McGuire, P. McMahon. 2012. Groundwater depletion and sustainability of irrigation in the US High Plains and Central Valley, *Proceedings of the National Academy of Sciences*, 109(24): 9320–9325.

Schnepf, R. 2012. *Agriculture-based Biofuels: Overview and Emerging Issues*, Washington, DC: Congressional Research Service, R41282.

Science for Energy Scenarios. 2014, 2016. www.science-and-energy.org (accessed May 2017).

Singh, R.K., H.R. Murthy, S.K. Gupta, A.K. Dikshit. 2012. An overview of sustainability assessment methodologies, *Ecological Indicators*, 15(1): 281–299.

Socolow, R., R. Hotinski, J.B. Greenblatt, S. Pacala. 2004. Solving the climate problem, technologies available to curb CO_2 emissions, *Environment*, 46(10): 8–19.

Spreng, D.T. 1988. *Net Energy Analysis and the Energy Requirements of Energy Systems*, New York: Praeger Publishing Company.

Statista, The Statistics Portal, http://www.statista.com (accessed February 2015).

Stern, N., 2006. *Stern Review: The Economics of Climate Change*, HM Treasury Cabinet Office, http://webarchive.nationalarchives.gov.uk/+/http:/www.hm-treasury.gov.uk/sternreview_index.htm.

Stuff. 2009. *Whispergen in Europe venture*, http://www.stuff.co.nz/business/242878/WhisperGen-in-Europe-venture (accessed November 2018).

Sullivan, J.L., L. Gaines. 2010. A review of battery life-cycle analysis: State of knowledge and critical needs, *Argonne National Laboratory*, ANL/ESD/10-7.

Sumner, D.A. 2009. Recent commodity price movements in historical perspective, *American Journal of Agricultural Economics*, 91(5): 1250–1256.

The Editors. (2011). Safety First, Fracking Second, *Scientific American.*

The Shift Project. 2010. http://www.tsp-data-portal.org/Breakdown-of-GHG-Emissions-by-Sector-and-Gas#tspQvChart (accessed May 2017).

TNS. 2015. The Natural Step, www.thenaturalstep.org/ (accessed October 2015).

Transport & Environment. 2017. Carmakers failing to hit their own goals for sales of electric cars, https://www.transportenvironment.org/sites/te/files/publications/2017_09_Carmakers_goals_EVs_report_I.pdf (accessed May 2018).

US Census Bureau. 2007. *Statistical Abstracts of the United States,* http://www.census.gov/prod/www/statistical-abstract.html (accessed 21 March 2008).

US DOE LLNL. 2014. Data based on DOE/EIA-0035(2014-03), Lawrence Livermore National Laboratory.

US DOE. 2006. Energy demands on water resources, Report to Congress, U.S., Washington, DC: Department of Energy.

US DOE. 2014. Greenhouse Gases, Regulated Emissions, and Energy Use in Transportation Model (GREET), Argonne National Laboratory, https://greet.es.anl.gov/ (accessed updated database March 2018).

US DOE/EPRI. 2013. *Electricity Storage Handbook in Collaboration with NRECA.* Albuquerque, NM: Sandia National Laboratories and Electric Power Research Institute.

US EIA. 2011. Primary Energy Production by Source, 1949–2010: U.S. Energy Information Administration, https://www.eia.gov/totalenergy/data/annual/showtext.php?t=ptb0103 (accessed updated data March 2018).

US EIA. 2012. Annual Energy Outlook 2012 with Projections to 2035, www.eia.gov/forecasts/aeo (accessed February 2019).

US EIA. 2017. Independent Statistics and Analysis, available at http://www.eia.gov/ (accessed February 2019).

US Environmental Protection Agency (EPA). 1990. 1990 Clean Air Act Amendment, https://www.epa.gov/clean-air-act-overview/1990-clean-air-act-amendment-summary (accessed July 2019).

US Environmental Protection Agency. 2005. National Emissions Inventory, http://www.epa.gov/hg/about.htm.

US Environmental Protection Agency (EPA). 2011. Mercury and Air Toxics Standards (MATS). https://www.epa.gov/mats/basic-information-about-mercury-and-air-toxics-standards (accessed July 2019).

US IEA. 2017a. OECD Library, Energy Technology RD&D Statistics, doi:10.1787/enetech-data-en.

UNGA. 2015. Transforming our world: The 2030 Agenda for Sustainable Development, United Nations General Assembly, A/RES/70/1.

United Nations. 2012. *World Population Prospects: The 2012 Revision,* Population Division of the Department of Economic and Social Affairs of the United Nations Secretariat, http://esa.un.org/unpd/wpp/index.htm.

USDA. 2015. Economic Research Service, Office of the Chief Economist, www.usda.gov/oce (accessed March 2018).

USDA. 2017. U.S. Drought Monitor, http://droughtmonitor.unl.edu/MapsAndData/MapArchive.aspx (accessed July 2017).

USGS. 2010. Mineral commodity summaries. Tech. Rep.; US Geological Survey; 2010, http://minerals.usgs.gov/minerals/pubs/mcs (accessed February 2015).

van Leeuwen, J.W.S. 2008. Nuclear Power Insights, Ceedata Consultancy, http://www.stormsmith.nl/insight-items.html (accessed March 2015).

Vidal, O. 2015. Personal discussion and research.

Voss, J.-P., A. Smith, J. Grin. 2009. Designing long-term policy: Rethinking transition management, *Policy Science,* 42: 275–302.

Wang, J., L. Feng, X. Tang, Y. Bentley, M. Hook. 2017. The implications of fossil fuel supply constraints on climate change projections: A supply-side analysis, *Futures*, 86: 58–72.

Wara, M. 2007. Is the global carbon market working? *Nature*, 445(7128): 595–596.

WCED. 1987. *Our Common Future*, Oxford, UK: Oxford University Press.

WEC. 2013. World Energy Council, www.worldenergy.org (accessed May 2017).

WFEO. 2015. World Federation of Engineering Organizations, *Engineers and Engineering Turning the Words of the COP-21 Agreement into Climate Action*, www.wfeo.org.

WFS. 2008. World Future Society Millennium Project, http://millennium-project.org/millennium/scenarios/energy-scenarios.html (accessed March 2018).

Winston, A.S. 2014. *The Big Pivot, Radically Practical Strategies for a Hotter, Scarcer, and More Open World*, Boston, MA: Harvard Business Review Press.

WIPG. 2015. Western Area Power Grid, https://www.wecc.biz/Reliability/2016%20SOTI%20Final.pdf (accessed 3 November 2018).

World Bank. 2003. *Development Data and Statistics*, www.worldbank.org/data (accessed May 2007).

World Energy Council. 2016. *World Energy Resources Reports*, www.worldenergy.org (accessed March 2018).

Worldwatch Institute. 1984–2016. *State of the World*, www.worldwatch.org/bookstore/state-of-the-world (accessed January 2017).

Wu, M., M. Mintz, M. Wang, S. Arora. 2009. *Consumptive Water Use in the Production of Ethanol and Petroleum Gasoline*, Lemont, IL: Argonne National Laboratory.

Yoon, M. et al. 2008. Calcium as the superior coating metal in functionalization of carbon fullerenes for high-capacity hydrogen storage, *Physical Review Letters*, 100: 206806.

Index

Note: Page numbers in italic and bold refer to figures and tables, respectively.

Printed in Great Britain
by Amazon